VLSI Electronics
Microstructure Science

Volume 16

Lithography for VLSI

VLSI Electronics
Microstructure Science

A Treatise Edited by

Norman G. Einspruch

VLSI Electronics Microstructure Science

Volume 16

Lithography for VLSI

Edited by

Norman G. Einspruch

College of Engineering
University of Miami
Coral Gables, Florida

R. K. Watts

Bell Laboratories
Murray Hill, New Jersey

1987

ACADEMIC PRESS, INC.
Harcourt Brace Jovanovich, Publishers

Orlando   San Diego   New York   Austin
Boston   London   Sydney   Tokyo   Toronto

ACADEMIC PRESS, INC.
Orlando, Florida 32887

United Kingdom Edition published by
ACADEMIC PRESS INC. (LONDON) LTD.
24–28 Oval Road, London NW1 7DX

Library of Congress Cataloging in Publication Data

Lithography for VLSI.

(VLSI electronics; v. 16)
Includes bibliographies and index.
1. Integrated circuits—Very large scale integration.
2. Printed circuits. 3. Lithography. I. Einspruch,
Norman G. II. Watts, R. K. (Roderick K.) Date
III. Series.
TK7874.V56 vol. 16   621.381'73 s   86-32156
ISBN 0—12—234116—3 (alk. paper) [621.381'73]

PRINTED IN THE UNITED STATES OF AMERICA

87 88 89 90   9 8 7 6 5 4 3 2 1

Contents

Chapter 1   **Optical Lithography**
R. K. Watts

Chapter 2   **Lumped Parameter Model for Optical Lithography**
Ron Hershel and Chris A. Mack

Chapter 3   **The Evolution of Electron-Beam Pattern Generators
for Integrated Circuit Masks at AT&T Bell
Laboratories**
D. S. Alles and M. G. R. Thomson

Chapter 4   **Electron Resist Process Modeling**

N. Eib, D. Kyser, and R. Pyle

Chapter 5   **Ion-Beam Lithography**

Benjamin M. Siegel

Chapter 6   **Alignment Techniques in Optical and X-Ray Lithography**

M. Feldman

Chapter 7   **Metrology in Microlithography**

D. Nyyssonen

Contents

Chapter 8   **Electrical Measurements for Characterizing Lithography**
Christopher P. Ausschnitt

List of Contributors

Numbers in parentheses indicate the pages on which the authors' contributions begin.

D. S. Alles (57), AT&T Bell Laboratories, 600 Mountain Avenue, Murray Hill, New Jersey 07974

C. P. Ausschnitt* (319), GCA Corporation, IC Systems Division, Bedford, Massachusetts 01730

N. Eib (103), General Technology Division, IBM Corporation, Hopewell Junction, New York 12533

M. Feldman (229), AT&T Bell Laboratories, 600 Mountain Avenue, Murray Hill, New Jersey 07974

Ron Hershel (19), Hershel Consulting Co., 1440 N. Albany, Albany, Oregon 97321

D. Kyser (103) Philips Research Laboratories, Cignetics Corporation, 811 E. Arquez Avenue, Sunnyvale, California 94086

Chris A. Mack (19), Department of Defense, 9800 Savage Road, Fort Meade, Maryland 20755

D. Nyyssonen† (265), National Bureau of Standards, Gaithersburg, Maryland 20899

R. Pyle (103), Data Systems Division, IBM Corporation, Hopewell Junction, New York 12533

B. M. Siegel (147), School of Applied and Engineering Physics, Cornell University, Ithaca, New York 14853

M. G. R. Thomson (57), AT&T Bell Laboratories, 600 Mountain Avenue, Murray Hill, New Jersey 07974

R. K. Watts (1), AT&T Bell Laboratories, 600 Mountain Avenue, Murray Hill, New Jersey 07974

* Present affiliation: Obelisk, Inc., 3 Clinton Street, Cambridge, Massachusetts 02139
† Present affiliation: CD Metrology, Inc., 18614 Mustard Seed Court, Germantown, Maryland 20755

Preface

Transistors will function at smaller dimensions than those in current production. A larger fraction of the cost of a modern production facility for integrated circuits is spent for lithographic systems than for any other type of process equipment. These two facts largely account for the great interest in the progress of lithography development.

Lithography is a vast multidisciplinary subject. Other books on resists are devoted to single divisions of the subject. This book, on the other hand, treats special topics from each branch of lithography, and also contains general discussion of some lithographic methods. It will be of interest to engineers, scientists, and technical managers in the semiconductor industry and to engineering and applied physics faculty and graduate students.

The first two chapters are devoted to optical lithography, the most important of the lithographic techniques. Chapter 1 includes tutorial material, and Chapter 2 describes a new model for photoresist exposure and development. Electron-beam lithography has found widespread application in photomask making. Resolution is limited to some extent by electron scattering. Chapter 3 covers electron lithography in general, and Chapter 4 covers electron resist exposure modeling, a topic relevent to all types of electron lithography systems.

Since the resolution limits of optical and x-ray lithography are similiar (0.2–0.3 μm linewidth), no chapter is devoted solely to x-ray lithography. Chapter 6 deals with mask/wafer alignment for x-ray proximity printing and for optical lithography.

Chapter 5 covers the fundamentals of ion-beam lithography. Highest resolution in resists is attained with proton beams. Mask-based proximity and projection proton lithography systems can offer both high resolution and high throughput. Chapters 7 and 8 on metrology cover characterization of lithography by measurements of various types.

Chapter 1

Optical Lithography

R. K. WATTS

AT&T Bell Laboratories
Murray Hill, New Jersey 07974

I. INTRODUCTION

The years since 1980 have been noteworthy for the continuing success of optical lithography as it has penetrated the 1-μm barrier. Competing technologies have been shooting at a moving target.

Optical lithography comprises the formation of images with visible or ultraviolet radiation in photoresist by contact, proximity, or projection printing. It is generally agreed that the limit of optical lithography lies in the range of 0.3 to 0.5 μm, although it is not known what tolerances will be manageable under production conditions. Clearly, optical lithography will continue to occupy the primary position in the immediate future.

II. OPTICAL RESISTS

Photoresists are of two types. A negative resist on exposure to light is made less soluble in a developer solution, while a positive resist becomes more soluble. Commercial negative resists, such as Kodak Microneg 747,

1

consist of two parts: a chemically inert polyisoprene rubber, which is the film-forming component, and a photoactive agent. The photoactive agent on exposure to light reacts with the rubber to form cross-links between rubber molecules, making the rubber less soluble in an organic developer solvent. The reactive species formed during illumination can react with oxygen and be rendered ineffective for cross-linking. Therefore, a nitrogen atmosphere is usually provided.

The developer solvent not only dissolves the unexposed resist. The exposed resist swells as the un-cross-linked molecules are dissolved away. The swelling distorts the pattern features and limits resolution to two to three times the initial film thickness.

Positive resists have two components: a resin and a photoactive compound dissolved in a solvent. The photoactive compound is a dissolution inhibitor. When it is destroyed by exposure to light, the resin becomes more soluble in an aqueous alkaline developer solution. The unexposed regions do not swell much in the developer solution, and so higher resolution is possible with positive resists. The development processes of projection printed images in positive resists have been modeled numerically [1]. It is an isotropic etching process. The sensitivity of most standard resists peaks in the spectral range $0.3-0.4$ μm. Two examples of commercially available positive resists are Shipley MP2400 and Hunt HPR-206.

The light intensity $I(\lambda, z)$ that is effective in exposing a volume element of resist at height $z < T$ above the substrate depends on the reflectivity of the substrate, the thickness T of the resist, and the convolution of the absorption spectrum of the resist with the spectrum of the incident light. The absorption spectra are fairly sharp, and only a small part of the lamp spectrum is effective. The refractive index of most resists is ~ 1.6, leading to an optical mismatch at the air–resist interface. If the substrate is highly reflective, the standing wave pattern set up has a node at the substrate and a peak amplitude that depends on resist thickness T. This variation in peak intensity with resist thickness becomes less with decreasing substrate reflectivity and increasing absorption by the resist. Since resist thickness varies at a step in substrate topology, the resulting difference in effective exposure leads to size variations in the resist image. Because of the high resolution of positive resists, the standing wave patterns are often well resolved, and the resulting corrugations are seen in the resist edge profile. Positive resists have replaced negative resists in most applications.

Photoresists are being developed for shorter wavelengths where higher resolution is possible. A few such deep ultraviolet resists are poly(methyl methacrylate) (PMMA), sensitive for $\lambda < 0.25$ μm, poly(butene sulfone), sensitive for $\lambda \lesssim 0.2$ μm, and MP2400, which is being used for $\lambda = 0.25$ μm. At these shorter wavelengths the radiation quantum is large enough to produce scission of the molecular chain.

Other properties of resists that are also quite important are good adhesion to the substrate and resistance to wet and dry etch processes. In general the commercially available optical resists are compatible with such processes. Adhesion of positive resists to certain types of substrate is less than for negative resists, and a thin silane adhesive layer is often first spun onto the substrate.

The unwanted variation of feature size in the resist image is due to many effects, some related to resist properties and resist processing and others to the exposure tool. These contributions can be to some extent separated by writing the edge acuity or slope of a feature edge $\partial T/\partial x$ as $\partial T/\partial x = (\partial T/\partial q)(\partial q/\partial x)$. The quantity x is measured perpendicular to the pattern edge and parallel to the plane of the substrate. The first factor is proportional to λ/q. The parameter q is the exposure dose. The second is proportional to the intensity gradient in the image, as shown, for example, in Figs. 2 and 3. (The parameters Q, γ, and W are defined later.)

Computer modeling of resist images is an important field of study. The reader will find an introduction to this subject in Chapter 10 of ref. 2 and in two of the chapters of this volume.

III. CONTACT AND PROXIMITY PRINTING

In contact printing a photomask is pressed against the resist-covered wafer with pressures typically in the range 0.05–0.3 atm, and exposure is by light of wavelength near 0.4 μm. Very high resolution ($\lesssim 1$-μm linewidth) is possible, but because of spatial nonuniformity of the contact, resolution may vary considerably across the wafer. To provide better contact over the whole wafer a thin (0.2-mm) flexible mask has been used; 0.4-μm lines have been imaged in 0.98-μm resist [3]. The contact produces defects in mask and wafer so that the mask, whether thick or thin, may have to be discarded after a short period of use. For example, Jones [4] reports mask defect densities increasing to 37 defects/cm^2 after 15 exposures from 13 defects/cm^2 after 5 exposures. Defects produced are pinholes in the chromium film, scratches, intrusions, and star fractures.

Nevertheless, contact printing continues to be widely used. Lin [5] has imaged 0.25-μm features in 1.8-μm-thick PMMA resist using 0.20- to 0.26-μm radiation. Quartz or Al_2O_3 mask substrates must be used to pass these shorter wavelengths, since the usual borosilicate glass strongly absorbs wavelengths less than 0.3 μm.

Proximity printing has the advantage of lack of contact between mask and wafer and thus longer mask life. Resolution is not as good as for projection printing, but printer cost is lower. Resolution is also less than for hard-contact printing.

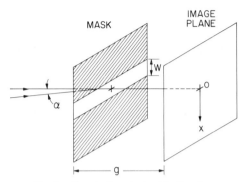

Fig. 1. Proximity printing idealization.

Figure 1 shows the optical problem in the schematic form of a mask with a long slit of width W separated from a parallel image plane (wafer) by a gap g. We assume that g and W are larger than the wavelength λ of the imaging light and that $\lambda \ll g < W^2/\lambda$, the region of Fresnel diffraction. Then the diffraction that forms the image of the slit is a function only of the particular combination of λ, W, and g, which we shall call the parameter Q, where

$$Q = W \sqrt{2/(\lambda g)}. \tag{1}$$

The diffraction patterns are well known. Some calculated by Skinner [6] are shown in Figs. 2 and 3 for $Q = 2$, 3, and 7 (solid curves). The limit $g < W^2/\lambda$ corresponds to $Q > \sqrt{2}$. The dashed rectangle shows the light intensity at the mask. The larger the value of Q, the more faithful is the image. We see that resolution becomes better at smaller gaps and shorter wavelengths.

Two other noteworthy features of the diffraction patterns are the ragged peaks and the slope near $x = W/2$. The peaks can be smoothed by the use of diverging rather than collimated light. The dotted curves are for light with divergence half-angle $\alpha = 1.5 \, \lambda/W$ rad. If α becomes large, the edge slope is reduced further, increasing linewidth control problems. Some smoothing also occurs because of the spread of wavelengths in the source, but this is a smaller effect. The illumination system must provide an apparent source size large enough to give a value of α, typically a few degrees, optimized for the smallest features to be printed. This is illustrated in Fig. 4. The mercury arc used as source is too small to yield the required α, so a scheme like that illustrated in Fig. 4c is used. The optical system must also minimize nonuniformity of intensity across the field; this may typically be 3–15%. Figure 5 shows the illumination system of a Canon proximity printer. The illumination is telecentric at the mask to prevent

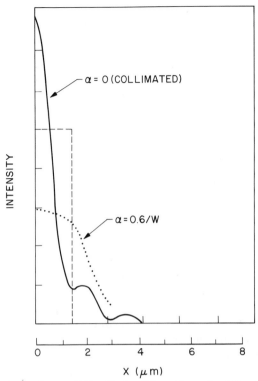

Fig. 2.   Image of slit with $Q = 2$ ($g = 10 \, \mu$m; $\lambda = 0.4 \, \mu$m).

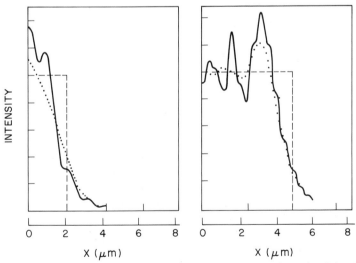

Fig. 3.   Image of slit with (left) $Q = 3$ and (right) $Q = 7$ ($g = 10 \, \mu$m; $\lambda = 0.4 \, \mu$m).

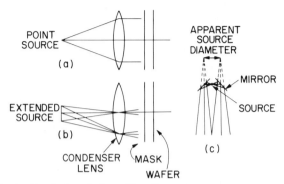

Fig. 4. Proximity printing with collimated and uncollimated light.

runout errors. With a mercury arc source the strong lines at 0.436, 0.405, and 0.365 μm provide exposure flux. The same printer is available with Xe–Hg source for enhanced output in the 0.2- to 0.3-μm spectral region. Using this source Nakane *et al.* [7] reported exposure times of 1 min (a very long exposure compared with standard resists) with PMMA resists and 2-μm resolution with $g = 10$–20 μm.

The edge of a feature in developed resist occurs at the position where the product of light intensity and exposure time It_e equals the resist threshold dose q_i. Linewidth control is in general more difficult in proximity printing than in contact printing. Thus, linewidth is influenced by variations in

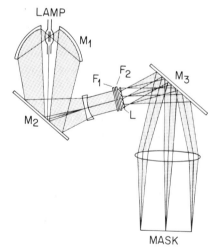

Fig. 5. Illumination system of Canon PLA 500F proximity printer (F, filter; M, mirror; L, fly's eye lens).

exposure intensity, gap g, and resist threshold. McGillis and Fehrs [8] measured linewidth as a function of exposure intensity I and mask–wafer gap. They found that the linewidth variations defined as printed width minus width of the line on the mask are proportional to $\ln I$ and to $\sqrt{g}$ and that for each value of g there is an optimum exposure intensity that minimizes the effect on linewidth of gap variation. The mask–wafer gap can vary across a wafer and from one wafer to another because of wafer and mask bowing and dirt particles between wafer and wafer chuck. This last problem is solved with a pin chuck. For exposure intensity near the optimum value, control of exposure to within 13% should provide line-width control of ± 0.5 μm, even for large variations in the gap g. Jones [4] has reported linewidth control to within ± 0.25 μm with a 50-μm gap.

Because of the sloping edges of the diffraction of the slit, the image of a mask object consisting of equally spaced parallel lines will lose definition as the spacing between lines decreases and the edge tails begin to overlap. If I_M is the peak intensity in the image (on the lines) and I_m is the minimum intensity in the image (between lines, where this quantity should be zero), the modulation of the image M_i is given by

$$M_i = (I_M - I_m)/(I_M + I_m). \qquad (2)$$

Modulation M_i is a function of linewidth, line spacing, gap, and wavelength. For 2.5-μm lines and spaces it depends on the mask–wafer gap in the following way, according to a calculation by Heim [9]:

g (μm)	M_i
5	0.62
10	0.50
20	0.25

For exposure of resist a ratio I_M/I_m of at least 4 is desirable. This corresponds to $M_i = 0.6$, a value that can be exceeded only at very small gaps. Further loss of modulation or contrast can be reduced by the use of Fe_2O_3 or Cr/Cr_2O_3 masks, which have much lower reflectivity than chromium and reduce scattered light under the opaque parts of the mask.

IV. PROJECTION PRINTING

Projection printing offers higher resolution than proximity printing together with large separation between mask and wafer. Four important performance parameters of a printer are resolution, level-to-level alignment accuracy, throughout, and depth of focus. The resolution of an

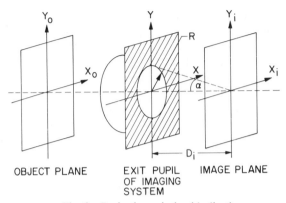

Fig. 6. Projection printing idealization.

optical imaging system of numerical aperture $\sin \alpha$ with light of wavelength λ is, according to Rayleigh's criterion, $0.61 \lambda/\sin \alpha$ — the separation of the barely resolved images of two point sources. The Rayleigh depth of focus is given by $\pm \lambda/(2 \sin^2 \alpha)$. At the limiting resolution of the system, the contrast in the image is uselessly small, however. We shall see how optical performance can be specified in a more meaningful way by means of the system transfer function.

We consider a general optical system (Fig. 6) with no aberrations. That is, a spherical wave diverging from a point in the object plane is converted to a spherical wave converging to a point in the image plane. The f number of the system is $F = D_i/2R$, and the numerical aperture is $\sin \alpha = R/\sqrt{(D_i^2 + R^2)} \approx R/D_i$.

Let us recall briefly some results from the theory of image formation [10] and show how the transfer functions differ for coherent and incoherent imaging systems. For coherent object illumination, all points in the object have wave amplitudes with fixed phase relationships, and all phases have the same time dependence. For example in Fig. 4a the phase varies across the plane at right in a way simply determined by the path lengths from the point source. If there is no phase correlation, on the other hand, and the phases vary independently from point to point across the object (Fig. 4b), the illumination is spatially incoherent. We consider first the coherent case. It is not necessary that the radiation be monochromatic, only that the wavelength spread be small compared with the average wavelength. The wavelength λ can stand for the average wavelength. The field amplitude $U_i(x_i, y_i)$ at point x_i, y_i in the image plane due to an amplitude distribution $U_o(x_o, y_o)$ in the object plane is given by

$$U_i(x_i, y_i) = \iint_{-\infty}^{\infty} K(x_i - x_o', y_i - y_o')U_o(x_o', y_o') \, dx_o' \, dy_o', \qquad (3)$$

where $x_o' = mx_o$ and m is the magnification of the system and K is the amplitude at x_i, y_i due to a point source at x_o, y_o. The amplitude K is given by the Fraunhofer diffraction pattern of the exit pupil $P(x, y)$,

$$K = \left(\frac{1}{\lambda D_i}\right)^2 \int\!\!\int_{-\infty}^{\infty} P(x, y) \exp\left\{\frac{-2\pi i}{\lambda D_i}[(x_i - x_o')x + (y_i - y_o')y]\right\} dx\, dy. \quad (4)$$

We shall be concerned only with a circular pupil function, as in Fig. 6. Then $P = 1$ for points x, y in the circle, and $P = 0$ outside the circle.

Let us consider a point source on axis, $x_o = y_o = 0$. Then Eq. (4) can be evaluated to give the Bessel function

$$K(x_i, y_i) = \frac{\pi}{(2\lambda F)^2}\left[\frac{2J_1(s)}{s}\right], \quad s = \left(\frac{\pi}{\lambda F}\right)\sqrt{x_i^2 + y_i^2}. \quad (5)$$

The quantity $[2J_1(s)/s]^2$, which is proportional to the intensity in the image of the point source, is the Airy pattern shown in Fig. 7. The first zero occurs at radius $x_i^{(0)} = 1.22\lambda F$. Eighty-five percent of the light is contained in the central spot and 7% in the first ring. Thus, Eq. (3) shows that the image is made up of a superposition of Airy patterns—one for each point

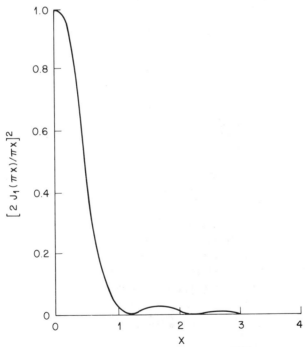

Fig. 7. Airy pattern. $X = (\pi/\lambda F)\sqrt{x_i^2 + y_i^2}$.

in the object. The smaller the aperture R, the larger is F and the lower is the resolution of the system, since each point is smeared into a circle of diameter $2.4\lambda F$.

The Fourier transforms of U_o, U_i, and K will be called G_o, G_i, and H respectively:

$$G(u, v) = \iint_{-\infty}^{\infty} U(x, y) \exp[2\pi i(ux + vy)] \, dx \, dy, \tag{6}$$

$$H(u, v) = \iint_{-\infty}^{\infty} K(x, y) \exp[2\pi i(ux + vy)] \, dx \, dy. \tag{7}$$

In Eqs. (6) and (7) G and U stand for G_o and U_o and G_i and U_i. The G_i and G_o are related by

$$G_i(u, v) = H(u, v)G_o(u, v), \tag{8}$$

and $H(u, v)$ is the frequency response function of the optical system. A component of the spatial frequency spectrum of the image is obtained by multiplying the corresponding component of the object spectrum by the value of H at that frequency. The function $H(u, v)$ is called the coherent transfer function. Taking the inverse transform

$$K(x, y) = \iint_{-\infty}^{\infty} H(u, v) \exp[-2\pi i(ux + vy)] \, du \, dv \tag{9}$$

and comparing with Eq. (4), we see that

$$H(u, v) = P(u\lambda D_i, v\lambda D_i). \tag{10}$$

The coherent transfer function has the same shape as the pupil function. For the circular aperture,

$$\begin{aligned} H(u, v) &= 1 \quad \text{if} \quad \sqrt{u^2 + v^2} \leq R/\lambda D_i \\ &= 0 \quad \text{if} \quad \sqrt{u^2 + v^2} > R/\lambda D_i. \end{aligned} \tag{11}$$

This transfer function is shown in Fig. 8. It has the value 1 for frequencies up to the cutoff frequency $u_m/2$, where $u_m = 2R/\lambda D_i = 1/\lambda F$.

Now let us consider spatially incoherent illumination where the phases vary independently across the object. The intensity $I_i(x_i, y_i)$ is a time average $U_i(x_i, y_i, t)U_i^*(x_i, y_i, t)$ and for incoherent light is given by

$$I_i(x_i, y_i) = \iint_{-\infty}^{\infty} I_o(x_o', y_o')|K(x_i - x_o', y_i - y_o')|^2 \, dx_o' \, dy_o'. \tag{12}$$

[For the coherent case it is given by the squared magnitude of the right side

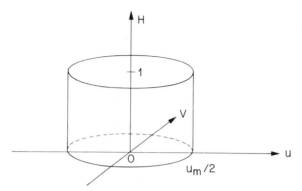

Fig. 8. Coherent or optical transfer function for a round pupil.

of Eq. (3).] If $\mathcal{G}_o$, $\mathcal{G}_i$, and $\mathcal{H}_i$ are Fourier transforms of object and image intensities and of $|K|^2$,

$$\mathcal{G}(u, v) = \iint_{-\infty}^{\infty} I(x, y) \exp[2\pi i(ux + vy)] \, dx \, dy, \tag{13}$$

$$\mathcal{H}(u, v) = \iint_{-\infty}^{\infty} |K(x, y)|^2 \exp[2\pi i(ux + vy)] \, dx \, dy, \tag{14}$$

then $\mathcal{G}_o$ and $\mathcal{G}_i$ are related linearly by

$$\mathcal{G}_i(u, v) = \mathcal{H}(u, v)\mathcal{G}_o(u, v).$$

The frequency response function $\mathcal{H}$ is called the optical transfer function, and $|\mathcal{H}|$ is called the modulation transfer function (MTF). Now, $|\mathcal{H}|$ is related to the pupil function in a more complicated way than before:

$$\mathcal{H}(u, v) = \left(\frac{1}{\lambda D_i}\right)^2 \iint_{-\infty}^{\infty} P\left(\xi + \frac{u\lambda D_i}{2}, \eta + \frac{v\lambda D_i}{2}\right)$$
$$\times P^*(\xi - u\lambda D_i/2, \eta - v\lambda D_i/2) \, d\xi \, d\eta. \tag{15}$$

In this case the transfer function is given by the overlap of two pupils displaced from one another by the amount $u\lambda D_i$. Figure 9 shows the case of the circular pupil, the line joining the pupil centers being taken as one axis. As u increases, the circles move apart and the transfer function decreases. By calculation of the shaded overlap area, $\mathcal{H}(u)$ is found to be

$$\mathcal{H}(u) = (2/\pi)[\cos^{-1}(u/u_m) - (u/u_m) \sqrt{1 - (u/u_m)^2}], \quad \text{for} \quad u \le u_m. \tag{16}$$

The limiting spatial frequency is twice as great as for coherent illumination, but image contrast or modulation is less at the lower frequencies. The transfer function describes how the system degrades the image of a sinusoi-

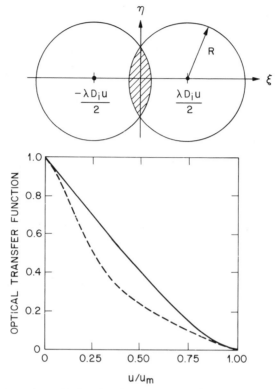

Fig. 9. Incoherent transfer function for a round pupil (——, perfect focus; ---, image plane displaced to give OPD = $\lambda/4$).

dal object grating. In general there is a reduction in modulation and a phase shift. The MTF at frequency u is the ratio of image modulation [see Eq. (2)] to object modulation for an object grating of periodicity u^{-1}:

$$|\mathcal{H}|(u) = M_i(u)/M_o(u) \tag{17}$$

King and Goldrick [11] describe MTF measurements for projection, proximity, and contact printers using a square wave bar target. The sine wave MTF can be calculated from the measured square wave response.

In modern printers the illumination is intermediate between the coherent and incoherent limits. The separation between points in the image that have correlated phases is neither infinite nor near zero. It can sometimes be varied by means of a diaphragm in the condenser illuminating the object. This is a familiar phenomenon to anyone who has used a microscope. As the aperture stop diaphragm is closed, the image looks "sharper," but interference rings appear around edges and the light level drops. This is the near-coherent case. Opening the diaphragm causes the rings to disappear and the image becomes brighter. The ratio S of the numerical aperture (NA) of the condenser to that of the projection optics is often used as a measure of coherence: $S = 0$ implies near coherence; if $S = 1$ the apertures are matched and the entrance pupil of the projection optics is filled to give near-coherent illumination. A further increase in S just causes more light scattering. Partial coherence has some advantages over incoherent illumination. The useful range, MTF > 0.6, is extended to higher spatial frequencies; edge gradients in the image become steeper; and the image is a little less sensitive to defocus. Figure 10 shows a transfer function for a lens with NA $= 0.28$ and partial coherence $S = 0.75$ for the wavelength 0.436 μm. This is compared with a curve for a silica lens with NA $= 0.38$ and incoherent illumination ($S = 1$); wavelength is 0.248 μm. Obviously, the second lens is capable of much higher resolution.

Since the transfer functions with round pupil are symmetric under rotation about the vertical axis, they can be written and displayed as functions of a single variable u as in Eqs. (16) and (17) and Figs. 8 and 9, but it is understood that u stands for $\sqrt{u^2 + v^2}$. This two-dimensional nature explains why a small contact hole with predominant spatial frequency components $u = v = u_o$ requires a different exposure from a long line of the same width with $u = u_o$, $v \approx 0$, for $|H(\sqrt{2}\ u_o)|$ is less than $|H(u_o)|$. If both types of feature occur on the same mask, both types of resist image will not have correct dimensions. In general for very small features representing high spatial frequencies, the required exposure depends on the

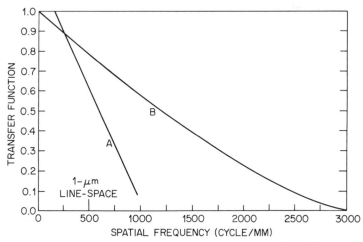

Fig. 10. Comparison of two transfer functions. (A) NA $= 0.28$; $S = 0.75$; $\lambda = 0.436$ μm. (B) NA $= 0.38$; $S = 1$; $\lambda = 0.248$ μm.

shape of the feature and proximity to other features, setting a practical resolution limit much less than u_m.

Aberrations can be treated by including a phase error $\phi(x, y)$ in the pupil function, $P(x, y) \rightarrow P(x, y) \exp[i\phi(x, y)]$. This reduces the magnitude of the transfer function, except in the coherent case, as can be seen from Eq. (10). Most printers have very nearly diffraction limited optics.

The focus error is a simple but important aberration. In Fig. 9 the dashed curve shows the effect of a displacement of the image plane or wafer from the focal plane by one Rayleigh unit $w = 2\lambda F^2$, corresponding to a phase error of $\pi/2$ at the edge of the pupil. Defocus affects different linewidths differently. Narasimham and Carter [12], using a Perkin Elmer printer, found that defocus of $3w$ changes the width of 7-μm lines by 2 μm and reduces the edge slope of the positive resist profile by 28°. Moritz, [13] using a similar printer and AZ1350J resist, found that a 3σ linewidth variation of ≤ 0.5 μm is achieved for defocus $\leq w$, some of the variation being due to illumination nonuniformity.

Some information on several projection printers is collected in Table I. These printers are of two types. The Perkin Elmer 600HT uses reflective optics, as illustrated in Fig. 11. A curved lamp source (not shown) illuminates an arc on the mask, and this is imaged onto the wafer with unity magnification ($m = 1$). In this way only a small zone of the spherical primary mirror is used, providing nearly diffraction limited imaging. Mask and wafer are swept through the arc to form an image of the whole mask. The other printers use refractive optics: high-quality lenses containing many elements. They project a 5$\times$ reduced image over the wafer.

As rough measures of resolution and depth of focus $l_{0.6}$, the linewidth of the equal line–space pattern for which the MTF of an incoherent system would have value 0.6 and w, the Rayleigh unit of defocus for an incoherent

TABLE I

Some Projection Printers

Printer	m	NA	Field (mm²)	$l_{0.6}$ (μm)	w (μm)	Align (μm)	Throughout (wafers/hr)
Perkin Elmer 600HT	1	0.16	150ϕ	1.9	±7.8	±0.35 (3σ)	100 (150 mm)
GCA 8000/1635	0.2	0.35	11 × 11	0.8	±1.5	±0.2 (3σ)	13 (150 mm)
GCA 8000/52529	0.2	0.29	18 × 18	1.1	±2.6	±0.2 (3σ)	37 (150 mm)
Nikon NSR 1505G4C	0.2	0.42	15 × 15	0.8	±1.2	±0.15 (3σ)	40 (125 mm)
Perkin Elmer SRA9535	0.2	0.35	17 × 17	0.9	±1.7	±0.15 (3σ)	50 (125 mm)
AT&T DUV[a]	0.2	0.38	10 × 10	0.5	±0.9	~±0.3	—

[a] Under development by AT&T Bell Laboratories.

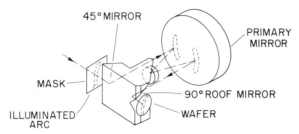

Fig. 11. Projection optics, Perkin Elmer printer.

system, are listed. Alignment tolerance and throughputs are those given by the manufacturer. If the whole image field cannot be filled with a large chip or several smaller ones and must be reduced in size, more steps are needed to cover the wafer, and throughout decreases. The very small depth of focus of the high-resolution optics requires close control of lens-to-wafer separation. Image distortion and magnification errors are generally small.

Since the size of the smallest feature resolvable with an optical system is proportional to λF, higher resolution is obtained by working at shorter wavelengths. There is more room for improvement here in projection printing than in proximity printing, where resolution is proportional to

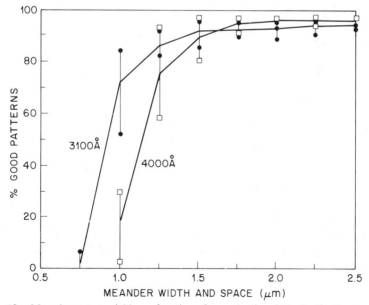

Fig. 12. Meander pattern yield as a function of meander linewidth. Perkin Elmer printer with $\lambda = 0.31$ and $0.4 \ \mu m$; $S = 0.86$.

$\lambda^{1/2}$. Alternatively, F can be reduced, but useful field size drops with decreasing F. In addition, depth of focus $\approx \lambda F^2$. Thus, it is better to reduce λ than F. The reduced depth of focus and degree of linewidth control accompanying higher resolution seem to represent an important practical limit. Some other problems encountered in the $\lambda = 0.2$- to 0.3-μm wavelength range are increased Rayleigh scattering ($\sim \lambda^{-4}$), greater difficulty in reducing aberrations, and a need for new resists. Reflective optics are more suitable; there are few optical glasses available for use in this wavelength range. Step-repeat exposure offers level-to-level registration precision independent of wafer size by separate alignment of each exposure field. The mask pattern dimensions are more convenient in the reduction projection systems than in those imaging with unity magnification.

Fig. 13. Pattern of 0.5 μm equal lines and spaces imaged in MP2400 resist with the AT&T DUV stepper of Table I. The pattern has been transferred through a tri-level resist structure.

Figure 12 shows yield of meander patterns as a function of meander linewidth as obtained by Bruning [14] with a modified Perkin Elmer printer at $\lambda = 0.31$ μm and $\lambda = 0.4$ μm. The feature sizes at which breakpoints in yield occur are roughly in the ratio of the two wavelengths, indicating diffraction-limited behavior at both wavelengths. The last entry in the table is an experimental stepper [15]. Source is an excimer laser operating at $\lambda = 0.248$ μm. The lens is all silica. Figure 13 shows a resist image produced by this system. A next-generation lens will provide the same resolution over a much larger field. Clearly, this machine is a portent of things to come. Wavelength can be reduced further to 0.19 μm without a change to reflective optics. From system considerations it seems that the limit of photolithography lies near 0.3 μm feature size.

REFERENCES

1. F. H. Dill, W. P. Hornberger, P. S. Hauge, and J. M. Shaw, *IEEE Trans. Electron Devices* **ED-22,** 445 (1975).
2. W. Fichtner, *In* "VLSI Technology" (S. M. Sze, ed.), Chap. 10.
3. H. I. Smith, N. Enfremow, and P. L. Kelley, *J. Electrochem. Soc.* **121,** 1503 (1974).
4. W. N. Jones, *Proc. Microelectron. Semin. Kodak Interface* **75,** 49 (1976).
5. B. J. Lin, *J. Vac. Sci. Technol.* **12,** 1317 (1975).
6. J. G. Skinner, *Proc. Microelectron. Semin. Kodak Interface* **73,** 53 (1974).
7. Y. Nakane, T. Tsumori, and T. Mifune, *Proc. Microelectron. Semin. Kodak Interface* **78,** 32 (1978).
8. D. A. McGillis and D. L. Fehrs, *IEEE Trans. Electron Devices* **ED-22,** 471 (1975).
9. R. C. Heim, *SPIE Semin. Proc.* **100,** 104 (1977).
10. M. Born and E. Wolf, *"Principles of Optics,"* 5th Ed. Pergammon, New York, 1975.
11. M. C. King and M. R. Goldrick, *Solid State Technol.* Feb., p. 37 (1977).
12. M. A. Narasimham and J. H. Carter, Jr., *Proc. SPIE* **135,** 2 (1978).
13. H. Moritz, *IEEE Trans. Electron Devices* **ED-26,** 705 (1979).
14. J. H. Bruning, *J. Vac. Sci. Technol.* **16,** 1925 (1979).
15. V. Pol, J. H. Bennewitz, G. C. Escher, M. Feldman, V. A. Firtion, T. E. Jewell, B. E. Wilcomb, and J. T. Clemens, *Proc. SPIE* **633,** 12 (1986).

Chapter 2

Lumped Parameter Model
for Optical Lithography

RON HERSHEL

Hershel Consulting Co.
Albany, Oregon 97321

CHRIS A. MACK

Department of Defense
Fort Meade, Maryland 20755

LIST OF SYMBOLS

D Resist thickness
D_{eff} Effective resist thickness
E Exposure energy
ϵ Natural log of exposure energy
E_0 Exposure energy needed to clear the resist
ϵ_0 Natural log of energy needed to clear the resist
γ Contrast (base e)
γ_D Developer contrast (base e)
γ_R Resist contrast (base e)
γ_T Total resist process contrast (base e)
γ_{10} Contrast (base 10)
I Relative image intensity
i Natural log of relative intensity
k Process-dependent resolution constant
λ Wavelength of exposing light (in vacuum)
NA   Numerical aperture of a lens
σ Coherence factor, or degree of coherence
T_r Relative resist thickness remaining after development
t_{dev} Development time
t_x Development time in the horizontal direction
t_z Development time in the vertical direction
τ Time needed to develop a unit thickness of resist
τ_0 Simplifying constant
r Development rate
r_0 Minimum development rate needed to clear resist

Dimensions

x Lateral dimension (in the wafer plane) measured from the mask center; also one-half
               of the dimension of the resist feature
z Vertical dimension measured into the resist from the top

I. INTRODUCTION

As the minimum feature size of integrated circuits becomes ever smaller, the importance of process and device modeling becomes more evident. This is never more true than for the process of optical lithography. Each small stride toward the theoretical limits of current lithography tools requires an exponential increase in our understanding of the process. Thus, process models have evolved to help study various aspects of the lithography process. Beginning with the pioneering work of Dill and co-workers at IBM [1], optical lithography modeling has advanced to the point of modeling projection and contact printing [2], multilayer resists, and contrast enhancement lithography [3]. Lithography models have been used to understand fundamental aspects of photoresist exposure and development [4,5] and to study the effects of often ignored processes such as photoresist prebaking [6].

The philosophy of these previous modeling studies has been to define each process as completely as possible. This requires the definition and evaluation of a large number of parameters. For example, the primary parameter model called the positive resist optical lithography (PROLITH) model requires more than 20 input parameters for a typical projection printing simulation. Determining the values of these parameters for each process to be modeled can be a significant task. This difficulty has impeded the use of primary parameter lithography models, especially in production environments. Here, production engineers tend to use a very few lumped parameters to describe their lithography process. Thus, there is a need for a lithography model that uses these lumped parameters to predict accurately effects that are important to the process engineer, such as exposure and focus latitude.

In this chapter, a lumped parameter model for optical lithography is defined, resulting in a direct relationship between the image intensity distribution and the critical dimension of the resulting pattern. The mathematical description of the resist process uses a simple photographic model relating development time to exposure, while the image simulation is derived from the standard optical parameters of the lithographic tool. The result is a fast and simple process simulator, based on test wafer results, which can accurately predict the behavior of resist critical dimensions over a sizable range of imaging parameters (numerical aperture, focus, etc.). With this model as a process controller, the photoengineer can quickly determine process latitude, that is, the range of exposure and focus that maintains feature sizes to within specified limits. The process latitude information can be used in selecting design rules and for reticle specification, as well as process control. This model also provides the framework for real-time optimization of the next generation of optical projection tools, which will offer variable resolution and field size.

II. BACKGROUND

A. Image Intensity

The image intensity is the image of the mask that is projected onto the wafer. Mathematically, the image intensity can be expressed as $I(x, y)$, where the wafer is in the $x-y$ plane. Typically, this function is normalized to the intensity in a large, clear area so that $I(x, y)$ is the relative image intensity and is dimensionless. The simplest type of mask features to study is an infinitely long line or space so that the image is one-dimensional [e.g., $I(x)$]. Methods for calculating the image intensity for typical projection printers have been reported [7,8], and an example of the resulting image is shown in Fig. 1.

There are three regions of interest in the image profile, as indicated in Fig. 1. The tail region is made up of the tail of the image plus any background intensity (caused, for example, by scattering and reflections from the lens surfaces). A typical background intensity may range from 1 to 4%. The tail controls photoresist erosion in the nominally unexposed regions (this is often called resist thinning). This may or may not be of importance, depending on the process. The second and most important part of the image profile is the toe. This is the region of the image near the mask edge. It is this region that controls the slope of the resist sidewall and, as we shall see, the exposure latitude. The center of the image, and in particular the value of the peak intensity, is important in determining the exposure energy needed to clear the feature in question relative to the energy needed for a large feature (which will have a peak intensity of 1, by definition).

There is another way to view the image intensity that yields useful information and insight. Figure 2 shows a plot of $-\ln\{I(x)\}$. This type of plot indicates an interesting feature of the image intensity: The toe region is nearly linear in log space. This can be found to be true not only for small

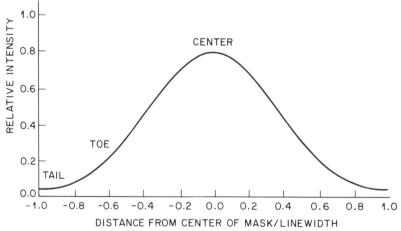

Fig. 1. Image intensity distribution for a typical step-and-repeat type projection printer (1.0-μm space; 2.0-μm pitch; NA = 0.28; $\sigma = 0.7$; $\lambda = 436$ nm).

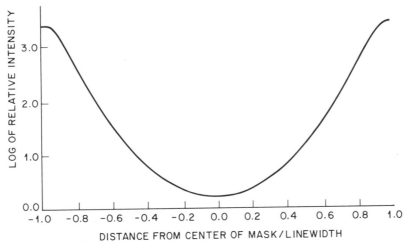

Fig. 2. Log-image plot of a typical image intensity distribution.

features, but for larger features as well (see Appendix A). Thus, the toe of an image profile is exponential. This fact will become very important in determining the effect of the image on exposure latitude.

A more complete description of the parameters needed to determine the shape of the image intensity distribution is given in Appendix A in the form of a short tutorial. Also given in the appendix are the effects of varying these parameters on the image.

B. Exposure and Focus Latitude

A photolithographic process can be defined by the functional relationship between the dimension of a critical feature (CD) and two process variables: focus and exposure. Thus, one of the most important curves in photolithography is the CD versus exposure energy curve (also called the exposure latitude curve). A typical example of such a curve is shown in Fig. 3. The ability to control the size of a critical feature is related to the slope of this curve. In the vicinity of the nominal feature size, one can see that the slope changes very rapidly. Thus, it is difficult to characterize dimensional control using a single value of the slope. (One should note that these curves are useless for comparison purposes unless plotted relative to the nominal exposure energy.)

Again, we will find it useful to plot these curves on a log-exposure scale, as in Fig. 4. One can see that there are regions of the CD curve that are nearly linear. Thus, the linewidth of a feature responds to the logarithm of

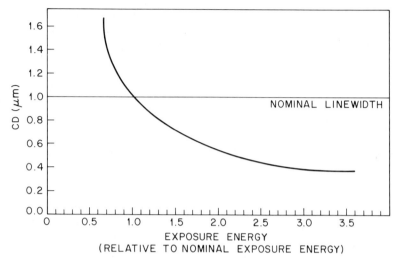

Fig. 3. Typical CD versus exposure curve for a 1.0-μm line.

the exposure energy over some range. The slope of this curve will provide a meaningful measure of the exposure latitude. The observed log-linear relationship is extremely important in understanding the behavior of photoresists and will form the basis of the lumped parameter model given below.

From curves such as those in Fig. 3 or 4, process latitude can be deter-

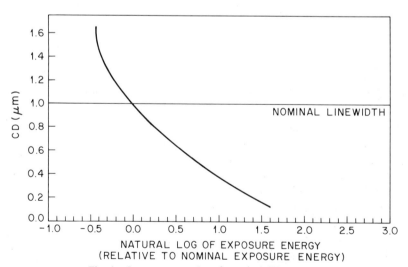

Fig. 4. Log-exposure plot of a typical CD curve.

mined as a function of exposure variations. These exposure variations include dose errors, illumination nonuniformity, changes in resist sensitivity and thickness, wafer reflectivity, etc. It is important to note that exposure errors generally vary as a percentage of the nominal exposure energy. For example, a 10% illumination nonuniformity will result in a 10% exposure error regardless of the exposure time. Thus, the most useful way of expressing exposure is relative to the nominal exposure energy.

The second important variable is focus. Focus errors result from auto-focusing errors, wafer flatness, topography, and lens aberrations. The exposure latitude curve for a defocused image is shown in Fig. 5. As can be easily seen, the exposure latitude decreases greatly with defocus. Thus, the two important process variables, focus and exposure, are not independent. As an example, consider a linewidth variation specification of ± 10%. For the perfect-focus case, one could determine from the exposure latitude curve the allowable exposure variation that would keep the linewidth "in spec." Following the same procedure for the defocused case, one would see a marked decrease in the allowable exposure error. In fact, by plotting the maximum exposure error as a function of defocus distance, one will define a focus–exposure window called the "process volume." An example of such a plot is shown in Fig. 6.

As an example of the use of the focus–exposure process volume, consider a printer that is known to have focus variations of ± 1 μm and a wafer with topography and flatness variations of ± 1 μm, giving a total focus error of ± 2 μm. This corresponds to a maximum exposure error of − 7% and

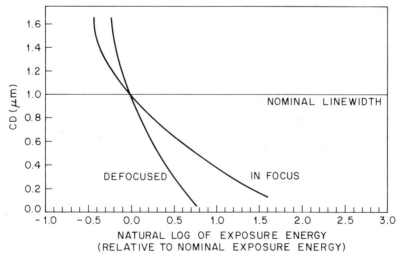

Fig. 5. Exposure latitude curves for nominally focused and defocused images.

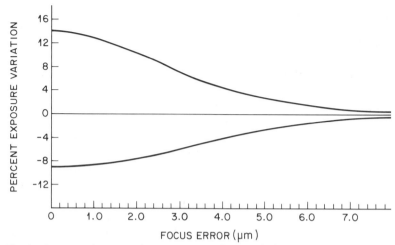

Fig. 6. Process volume: maximum exposure–focus variations that keep linewidth variations within specified limits.

+10% (lightly shaded rectangle in Fig. 7). If, however, the total focal error is ±4 μm, the maximum allowable exposure error is reduced to ±4% (darkly shaded rectangle in Fig. 7). Thus, it is impossible to define either exposure latitude or focus latitude independently.

A simple model will now be defined that predicts the effects of these focus and exposure errors on critical dimensions.

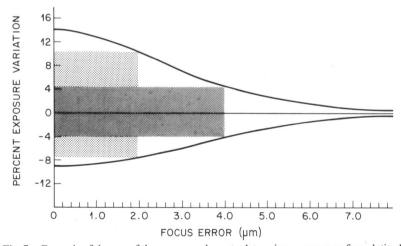

Fig. 7. Example of the use of the process volume to determine exposure or focus latitude.

III. LUMPED PARAMETER MODEL

A. Derivation

A complete derivation of the lumped parameter model in two dimensions is given in Appendix B. Here, an outline of that derivation will be given; a few of the important steps will be pointed out, and all the assumptions stated.

The lumped parameter model is based on a model for the development process, which in turn is based on the characteristic curve (also called the contrast curve) of a photoresist. Following the discussion above, logarithmic definitions of the image intensity and exposure energy will be made in the hope of deriving a formalism for the observed log-linear relationship of the CD curve. Let E be the nominal exposure energy and $I(x)$ the image intensity. It is clear that the exposure energy as a function of x is

$$E(x) = EI(x). \tag{1}$$

Defining logarithmic versions of these quantities, one obtains

$$\epsilon(x) = \ln[E(x)],$$
$$\epsilon = \ln[E], \tag{2}$$
$$i(x) = \ln[I(x)].$$

Thus, Eq. (1) becomes

$$\epsilon(x) = \epsilon + i(x). \tag{3}$$

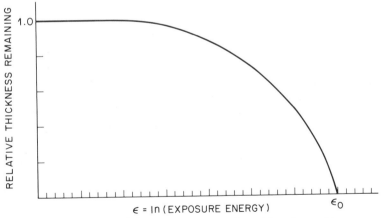

Fig. 8. Characteristic or contrast curve of a positive photoresist.

These logarithmic definitions will also be useful in dealing with the characteristic curve of a photoresist, which uses $\log_{10} E$ as the abscissa.

Shown in Fig. 8 is a photoresist characteristic curve that uses the natural log instead of the conventional $\log_{10} E$. This curve relates resist thickness remaining after development to exposure energy. By examining the curve, one might expect that a reasonable fit to this curve can be obtained using an exponential function. In particular, the relative thickness remaining T_r can be modeled as

$$T_r = 1 - e^{\gamma(\epsilon - \epsilon_0)}, \tag{4}$$

where ϵ_0 is the energy required to just clear the photoresist in the allotted development time. The use of the letter γ for the constant in the exponential is not arbitrary. It is easy to show that the slope of the curve given by Eq. (4) at $\epsilon = \epsilon_0$ is just $-\gamma$. Thus, γ is related to the conventional base 10 contrast of the resist process, which we shall call γ_{10}, by

$$\gamma_{10} = 2.303\gamma. \tag{5}$$

If the development rate is assumed constant through the resist (i.e., if resist absorption and standing waves are ignored), the relative thickness remaining can be related to the development rate by

$$T_r = 1 - r(E)t_{dev}/D = 1 - r/r_0, \tag{6}$$

where D is the resist thickness, t_{dev} the development time, r the development rate as a function of exposure energy, and r_0 the development rate needed to just clear the resist in the allotted development time. Comparing Eqs. (4) and (6), one can see that

$$r(x) = r_0 e^{\gamma(\epsilon(x) - \epsilon_0)}. \tag{7}$$

Equation (7) is a simple model relating development rate to exposure energy based on the common characteristic curve of a photoresist. In order to use this expression, we shall develop a phenomenological explanation for the development process. This explanation is based on the assumption that development occurs in two steps: a vertical development to a depth z, followed by a lateral development to position x (measured from the center of the mask feature). This type of segmented development process is illustrated in Fig. 9. A development ray, which traces out the path of development, starts at the point $(x_0, 0)$ and proceeds vertically until a depth z is reached such that the resist to the side of the ray has been exposed more than the resist below the ray. At this point the development will begin horizontally. The time needed to develop in both vertical and horizontal directions, t_z and t_x, respectively, can be computed from Eq. (7).

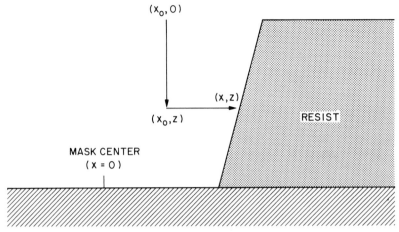

Fig. 9. Segmented development concept.

The development time per unit thickness is simply the reciprocal of the development rate,

$$\tau(x) = \tau_0 e^{-\gamma \epsilon(x)}, \tag{8}$$

where

$$\tau_0 = e^{\gamma \epsilon_0}/r_0.$$

The time needed to develop horizontally from x_0 to x is given by

$$t_x = \tau_0 e^{-\gamma \epsilon} \int_{x_0}^{x} e^{-\gamma i(x')} \, dx'. \tag{9}$$

The total development time is fixed at t_{dev}, so that

$$t_{\text{dev}} = t_z + \tau_0 e^{-\gamma \epsilon} \int_{x_0}^{x} e^{-\gamma i(x')} \, dx'. \tag{10}$$

Equation (10) can be used to derive some interesting properties of the resist profile. For example, how would a small change in exposure energy, $d\epsilon_0$, change the final position of the development ray? One can see that a change in exposure energy will not change the depth z at which the ray turns to a horizontal direction. The effect on the position x can be derived by differentiating Eq. (10), with the result

$$\left. \frac{dx}{d\epsilon} \right|_z = \frac{\gamma t_{\text{dev}}}{\tau(x)} = \gamma t_{\text{dev}} \, r(x). \tag{11}$$

· Since the x coordinate is simply one-half the width of the feature (measured at a depth z), Eq. (11) defines the change in CD with exposure energy. Two forms of this equation will be given. The first can be obtained by taking the log of both sides to give

$$i(x) = \frac{1}{\gamma} \ln \left(\frac{dx}{d\epsilon} \right) - \epsilon(x) + \frac{1}{\gamma} \ln \left(\frac{\tau_0}{\gamma t_{dev}} \right). \tag{12}$$

Equation (12) is the log-linear form of the lumped parameter model and relates the slope of a CD versus log-exposure curve to the image intensity. Using this expression, one can use CD versus exposure data to calculate the image intensity distribution $i(x)$.

A second form of the lumped parameter model can also be obtained from Eq. (11), giving

$$\epsilon(x) = \epsilon(0) + \frac{1}{\gamma} \ln \left[1 + \frac{1}{D} \int_0^x \left(\frac{I(x)}{I(0)} \right)^{-\gamma} dx \right], \tag{13}$$

where $\epsilon(x)$ is the (log) energy needed to expose a space of width $2x$. The term $\epsilon(0)$ represents the exposure energy needed to give a CD of 0, that is, to just clear the resist. Equation (13) is the integral form of the lumped parameter model. Using this equation, one can generate a normalized CD versus exposure curve by knowing the image intensity $I(x)$, the resist thickness D, and the (base e) contrast γ.

Examining Eq. (12), one can make an observation that was hinted at in the previous section. If the slope of the CD versus log-exposure curve (i.e., $dx/d\epsilon$) is constant, there is a direct relationship between $-i(x)$ and $\epsilon(x)$. That is, the log-image curve, as shown in Fig. 2, is controlling the shape of the CD versus log-exposure curve, as shown in Fig. 4, over the region where $dx/d\epsilon$ is constant. Also, if the contrast of the resist process is very large (i.e., approaching infinity), $\epsilon(x)$ becomes $-i(x)$. In such a case, the CD curve is image dominated and represents the best possible exposure latitude.

B. Contrast

It is obvious from Eq. (13) that the contrast γ is a very important parameter in determining the exposure latitude. Therefore, some comments on the definition and derivation of the contrast used in the lumped parameter model are in order. The simple development model given by Eq. (7) is derived by assuming a nonabsorbing resist and fitting the contrast curve of this resist to a simple equation. Although later the effects of absorption are added to our development model (see Appendix B), the parameter γ still applies to this idealized nonabsorbing resist. It is known that the effect of resist absorption is to decrease the contrast (dyed resists have very low contrast values). However, the value of γ used in the lumped parameter model still applies to a nonabsorbing resist.

It has been proposed [5] that the contrast of a resist system can be broken up into two components, developer contrast γ_D and resist contrast γ_R, due to absorption. The developer contrast is the contrast of the resist system with no absorption (i.e., infinite resist contrast). The resist contrast is the contrast of the system assuming a perfect developer (i.e., an infinite developer contrast). The overall contrast γ_T is then given by

$$\gamma_T = (1/\gamma_D + 1/\gamma_R)^{-1}. \tag{14}$$

Thus, the γ used in Eq. (13) is not the conventional photoresist contrast, but the developer contrast. Throughout the rest of this chapter, the term *contrast* will refer to the developer contrast.

C. Validity of Assumptions

The lumped parameter model represented by Eq. (12) is based on two assumptions. First, we assume that a photoresist characteristic curve can be fit by Eq. (4). If the fit is good, the development rate model given by Eq. (7) will be accurate. Obviously, these equations assume that the energy within the resist $\epsilon(x)$ is less than ϵ_0. Thus, energies greater than ϵ_0 are not modeled, and one must be careful when using Eq. (12) to keep this in mind. The limitations of this restriction are not severe for this application, however, since in the vicinity of the final resist sidewall the energy must necessarily be less than ϵ_0.

The second assumption is that the development process is segmented into vertical and horizontal steps. Obviously, the validity of this assumption would be difficult to determine experimentally. It is, however, quite easily accomplished with a primary parameter lithography process model such as PROLITH. The development model used by PROLITH is based on a kinetic analysis of a proposed reaction mechanism [5]. The method used to compute the resist profile is a string algorithm that computes the position of the resist–developer interface as the development process proceeds. Thus, by following a particular point on the interface a development ray is traced.

Figure 10 shows the results of such ray tracing for the case of resist exposure on a nonreflecting substrate. One can see that for this case the development process is continuous, not segmented. This result would be expected since the change in exposure energy vertically is quite gradual

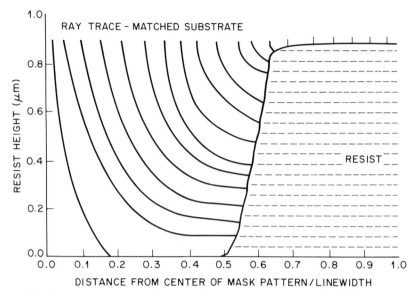

Fig. 10. Development rays for a matched substrate as predicted by PROLITH.

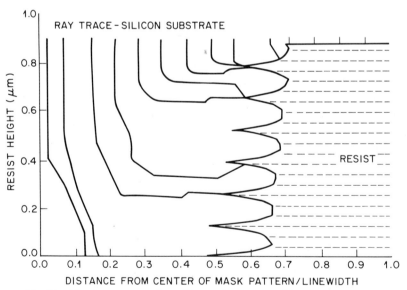

Fig. 11. Development rays for a silicon substrate as predicted by PROLITH.

(due only to absorption). When a reflecting substrate is used, however, there are dramatic differences in exposure energies over small vertical distances (due to standing waves). Thus, it seems likely that the exposure condition given in the above analysis will be met at some point and the development process will occur in sequential vertical and horizontal steps. As can be seen in Fig. 11, the development rays change direction quite sharply when standing waves are present.

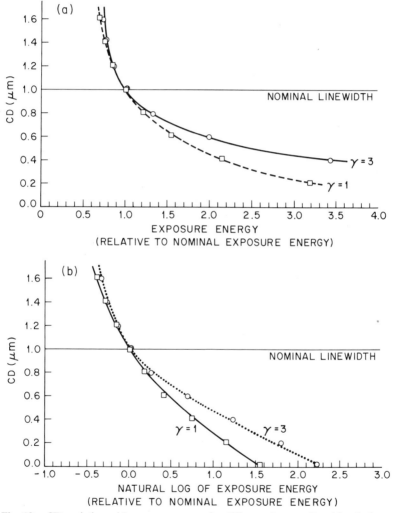

Fig. 12. CD variation with exposure energy for different γ's as predicted by the lumped parameter model (0.9-μm resist, 1.0-μm lines and spaces). (a) CD versus exposure; (b) CD versus log exposure.

D. Using the Model

Equation (13) can now be used to generate exposure latitude curves (CD versus exposure energy). Figure 12a shows one such curve for 1.0-μm lines and spaces simulated using typical g-line projection printer parameters (NA = 0.28, $\sigma = 0.7$) and using 0.9 μm of resist. The image intensity distribution was simulated using PROLITH v1.2 and the resulting data

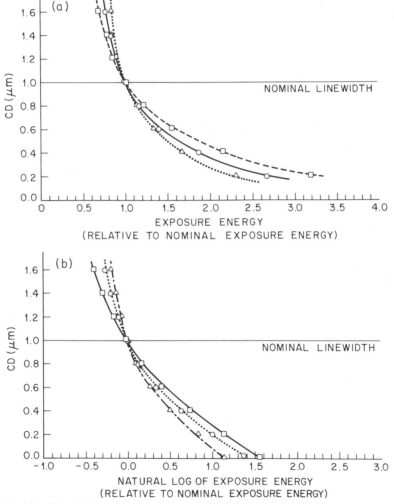

Fig. 13. CD variation with exposure energy for different resist thicknesses as predicted by the lumped parameter model ($\gamma = 1$, 1.0-μm lines and spaces). (a) CD versus exposure; (b) CD versus log exposure. □, $D = 0.9$ μm; O, $D = 1.4$ μm; △, $D = 2.0$ μm.

numerically integrated in Eq. (13). As can be seen, the resist contrast (γ) plays a critical role in determining the process latitude. As noted earlier, a more useful form of this curve is the CD versus log-exposure curve (Fig. 12b). Both types of plot are shown throughout the chapter for comparison purposes. Increasing resist thickness results in a loss of process latitude (Fig. 13). This effect is quite noticeable in a low-γ process, but for higher-

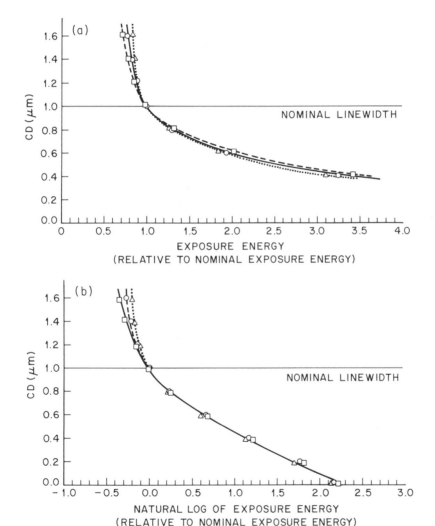

Fig. 14. CD variation with exposure energy for different resist thicknesses as predicted by the lumped parameter model ($\gamma = 3$, 1.0-μm lines and spaces). (a) CD versus exposure; (b) CD versus log exposure. □, $D = 0.9\ \mu$m; ○, $D = 1.4\ \mu$m; △, $D = 2.0\ \mu$m.

contrast resist systems, thickness has fewer effects on exposure latitude
(Fig. 14). Figure 15 illustrates the well-known fact that small features have
less exposure latitude than larger features. Finally, Fig. 16 shows how
defocus degrades exposure latitude.

All data in Figs. 12–16 were generated using the lumped parameter

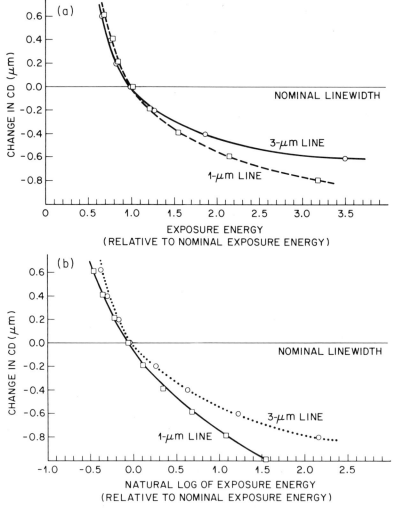

Fig. 15. CD variation with exposure energy for different feature sizes as predicted by the
lumped parameter model ($\gamma = 1$, 0.9-μm resist). (a) CD versus exposure; (b) CD versus log
exposure.

model. Some of the trends shown in these figures may be obvious to an experienced lithography engineer; some may not. In any case, these trends will be verified by comparing the lumped parameter model with experimental data.

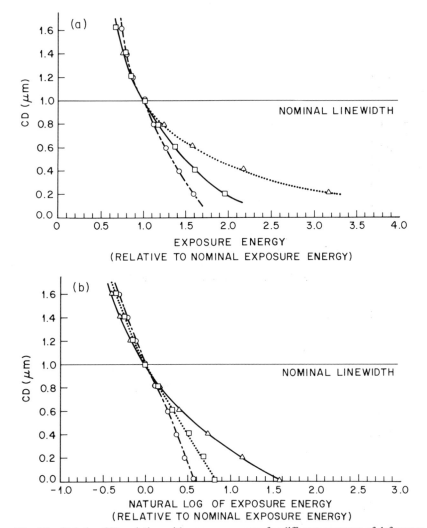

Fig. 16. Relative CD variation with exposure energy for different amounts of defocus as predicted by the lumped parameter model ($\gamma = 1$, 0.9-μm resist, 1.0-μm lines and spaces). (a) CD versus exposure; (b) CD versus log exposure. $\triangle$, no defocus; $\square$, 2-μm defocus; $\bigcirc$, 3-μm defocus.

IV. COMPARISON WITH EXPERIMENTAL DATA

The true test of any model is its capacity to describe adequately experimental data. In this case, the data are scanning electron microscope automatic linewidth measurements made for a variety of exposure energies repeatable to $\pm 0.02\ \mu$m and accurate to $\pm 2.5\%$. An Ultratech 1000 stepper

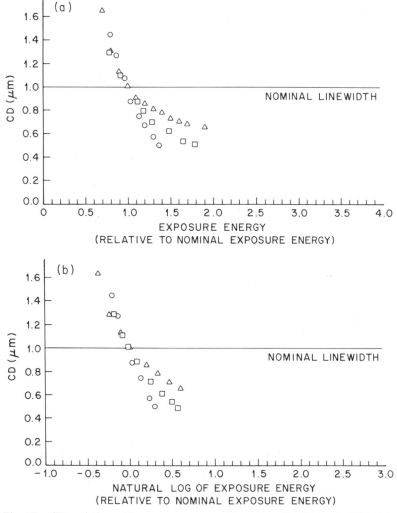

Fig. 17. CD variation with exposure energy—data taken for Ultratech 1000 (1.1-μm AZ1470 resist on silicon, 90-sec development, 1.0-μm lines and spaces). (a) CD versus exposure; (b) CD versus log exposure. $\triangle$, no defocus; $\square$, 2-μm defocus; $\bigcirc$, 3-μm defocus.

(NA = 0.315, $\sigma = 0.45$) was used to expose 1.1 μm of AZ1470 resist on silicon wafers. The wafers were developed for 90 sec in 5:1 AZ developer. Although the Ultratech uses broadband exposure in the range of 380 to 450 nm, for the purposes of calculating the image intensity a wavelength of 420 nm was used. Equal lines and spaces of 1.0 μm were imaged for five different focus distances, 0 (in perfect focus), ± 2 μm, and ± 3 μm, all accurate to ± 0.25 μm. The resulting data are shown in Fig. 17.

A. Lumped Parameter Estimation

According to Eq. (13), only one parameter, γ, can be varied in order to fit the lumped parameter model to experimental data. However, modeling studies using PROLITH to generate exposure latitude data for different substrates indicate that changes in substrate reflectivity modify the shape of the CD curves in the same way as changes in resist thickness.* Thus, there is physical significance in allowing the resist thickness D to be replaced by an effective resist thickness D_{eff} in order to account for substrate differences. Therefore, one must determine two parameters when fitting experimental data. Also, a background intensity for the image can be estimated based on the linewidth data.

The lumped parameters can now be easily determined. An excellent fit was obtained over the full range of CDs and for all five focus settings using $\gamma = 1.6$, $D_{eff} = 1.5$, and a background intensity of 2.5% (Fig. 18).

B. Process Interpolation

Given the excellent fit of the model to the wide range of exposure and focus data, one would expect that the model could then accurately predict the behavior of the process at other focus settings and reticle sizings. Thus, one could predict, for example, the maximum defocus that would keep the linewidth within ± 10% for a given exposure variation. Using the lumped parameters given above, a focus–exposure window, called the process volume, can be generated for a given linewidth specification. Focus and exposure errors within this volume will keep the linewidth within specifications. This type of plot is extremely important in showing the interdependence of exposure and focus errors on the resulting linewidth.

The lumped parameter model can also be used to explore the effects of mask biasing on process latitude. For example, a series of 1.0-μm lines and spaces can be modeled and the process volume determined. Similarly, a

---

* Modeling studies comparing the primary and lumped parameter models will be published elsewhere.

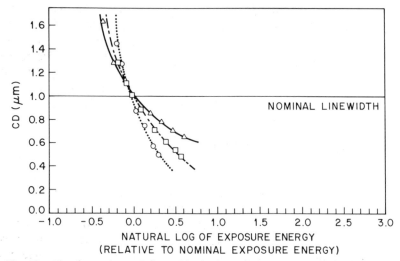

Fig. 18. Fit of experimental data with the lumped parameter model. All curves use $\gamma = 1.6$ and $D_{\text{eff}} = 1.5$ μm. Experimental data: $\triangle$, no defocus; $\square$, 2-μm defocus, $\bigcirc$, 3-μm defocus. Lumped parameter: ——, no defocus; – – –, 2-μm defocus; $\cdot\cdot\cdot$, 3-μm defocus.

1.1-μm line, 0.9-μm space combination (i.e., a 0.1-μm mask bias) can be modeled and the process volume determined for a desired linewidth of 1.0 μm. One can then see the advantage (or disadvantage) of mask biasing with respect to increased process latitude.

C. Process Extrapolation

The performance of the resist process can also be estimated for other printers (e.g., different numerical apertures). Thus, the trade-off between resolution and depth of focus can be determined for a specific resist process and for a particular linewidth specification. This type of analysis can be very useful when evaluating the purchase of a new lithographic tool. Also, the ability to predict the effects of a numerical aperture change on a particular resist process will be an essential part of the operation of variable numerical aperture tools, which are currently under development.

As a final note, CD versus exposure data can also be used to calculate the image profile $i(x)$ using Eq. (12). Thus, the lumped parameter model represents one of the few methods of determining the image profile experimentally. Using the infocus data of Fig. 17, Eq. (12) was applied and a predicted image profile was determined. Further, two more sets of CD versus exposure data were taken using 30- and 180-sec development times,

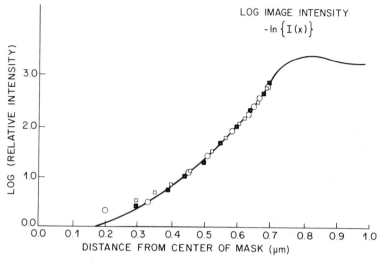

Fig. 19. Determination of the image profile based on three sets of measured CD versus exposure data. □, 180-sec development; ■, 90-sec development; ○, 30-sec development; ——, theoretical image.

and values for $i(x)$ were calculated. These values are shown in Fig. 19. The solid line represents the theoretical image profile as predicted by partial coherence theory. One can see that the fit is remarkable, with only slight deviation near the center of the image.

V. SUMMARY

A simple model has been presented that predicts the exposure latitude of a resist process for a given image profile. The resist process is governed by two lumped parameters: the resist process contrast and the effective resist thickness. Once these two parameters have been determined (using measured CD versus exposure data), the exposure latitude for any image profile can be predicted. Thus, the effect of defocus on exposure latitude can be quickly determined, allowing for the calculation of the process volume of the resist process. A knowledge of the process volume is absolutely essential in order to provide adequate process control for a high-resolution lithographic process. The ability to describe the process volume mathematically is the first step in the development of an automated photolithographic process control system.

APPENDIX A: IMAGE INTENSITY DISTRIBUTION

In order to use the lumped parameter model given in this chapter, one must first be able to predict the image intensity distribution $I(x)$ for a given set of parameters. The numerical details of such simulations have been given previously (e.g., [7,8]) and will not be repeated here. Instead, definitions of each of the important image parameters will be presented, and their effect on the image intensity distribution will be shown.

A typical step-and-repeat type of projection printer is an extremely complicated collection of optical elements whose function is to project an image of a mask or reticle onto a photoresist-coated wafer. As complicated as this system may be, its operation is that of a simple two-lens projector, as shown in Fig. A-1. Ultraviolet light emitted from the source (usually a high-pressure mercury arc lamp) passes through a filter, which makes the light essentially monochromatic. This light is then projected onto the mask by the condenser lens. The mask diffracts light, which is collected by the objective lens and imaged on the wafer.

One of the most important parameters of a lens in this system (or in any optical system) is the numerical aperture, as defined in Fig. A-2. In fact, the numerical apertures of the two lenses along with the wavelength of light used is a sufficient set of parameters to predict the performance of an ideal optical system of this type, assuming the wafer is in perfect focus. Another parameter, called the degree of coherence or the coherence factor σ, can be defined as

$$\sigma = \frac{\text{numerical aperture of condenser}}{\text{numerical aperture of objective}} \tag{A.1}$$

When σ is zero, the illumination is said to be coherent. When σ is infinite, the illumination is incoherent. Finite nonzero values of σ represent partially coherent illumination. Most projection systems used in lithography have coherence factors in the range 0.4–0.7.

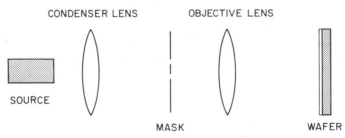

Fig. A-1.   Diagram of an optical projection system.

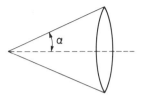

NA = sin α

Fig. A-2.   Definition of the numerical aperture (NA) of a lens.

As stated, this optical system projects an image of the mask onto the wafer. As with any imaging system, the wafer must be the proper distance from the objective lens in order for the image to be in focus. If the wafer is moved from the plane of perfect focus by a distance δ, the image is said to be defocused by this distance. Current high-resolution lithography tools can tolerate no more than a few micrometers of defocus before performance is degraded significantly.

Once the parameters of the optical system have been defined, the image intensity at the wafer plane can be determined for a given mask pattern. For simplicity, we shall deal with infinitely long lines or spaces (or both) so that the mask (and the image) is one-dimensional. For the case of equal lines and spaces of width 1.0 μm, the image intensity corresponding to a space is shown in Fig. A-3a. The parameters used in the simulation are for a typical g-line projection printer. As was discussed in the chapter, a plot of $-\ln\{I(x)\}$ (Fig. A-3b) is also an important representation of the image. Thus, all images will be shown in both types of plot.

Using Fig. A-3 as our baseline image, we shall now study the effects of varying each of the parameters of the projection system. Figure A-4 shows the effect of wavelength on the image. Obviously, lower wavelengths produce superior images. From Fig. A-5 one can see that higher numerical apertures (of the objective lens) also improve image quality. In fact, the resolution of a printer, that is, the smallest feature that can be printed, is proportional to wavelength divided by the numerical aperture. That is,

$$\text{Resolution} = k\lambda/\text{NA} \qquad (\text{A.2})$$

where k is some process-dependent constant. The theoretical limit is often said to be at $k = 0.6$, the Rayleigh resolution limit. The degree of coherence is also important in determining the performance of a projection printer, as seen in Fig. A-6. Finally, the degradation of the image due to defocus is quite evident in Fig. A-7.

Figures A-8 and A-9 show a very important feature of the image inten-

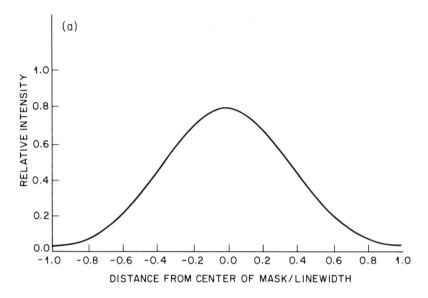

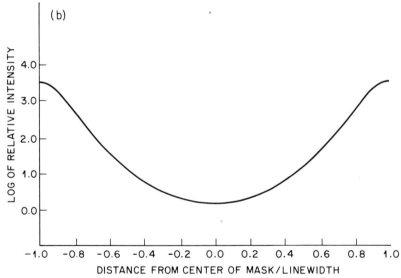

Fig. A-3. Image intensity distribution for a typical *g*-line stepper. (a) Conventional image plot; (b) log-image plot. 1.0-μm space, 2.0-μm pitch, NA = 0.28, σ = 0.7, λ = 436 nm; no defocus.

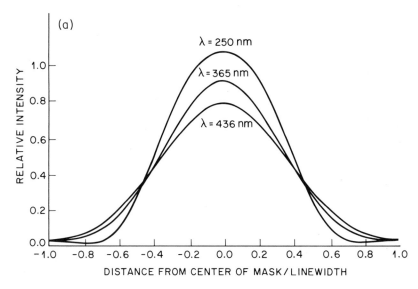

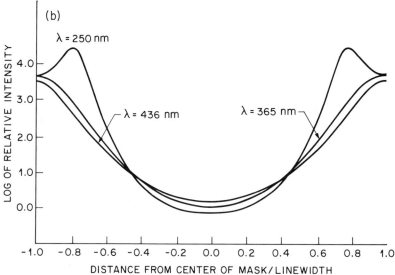

Fig. A-4. Variation of image intensity distribution with wavelength. (a) Conventional image plot; (b) log-image plot. 1.0-μm space, 2.0-μm pitch, NA = 0.28, σ = 0.7; no defocus.

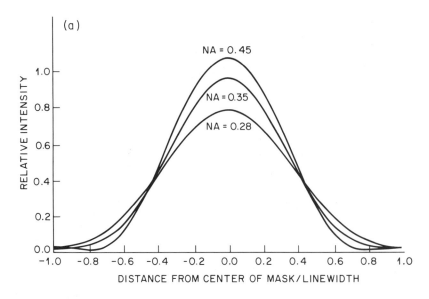

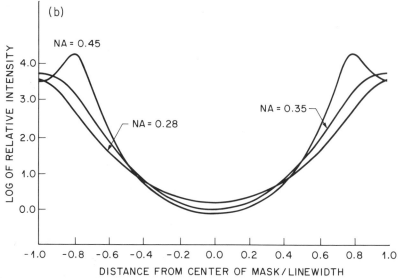

Fig. A-5. Variation of image intensity distribution with the numerical aperture of the objective lens. (a) Conventional image plot; (b) log-image plot. 1.0-μm space, 2.0-μm pitch, $\sigma = 0.7$, $\lambda = 436$ nm; no defocus.

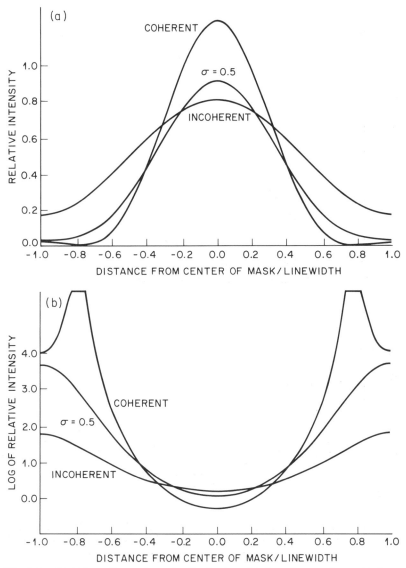

Fig. A-6. Variation of image intensity distribution with the coherence factor. (a) Conventional image plot; (b) log-image plot. 1.0-μm space, 2.0-μm pitch, NA = 0.28, λ = 436 nm; no defocus.

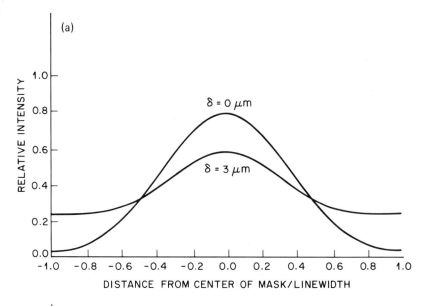

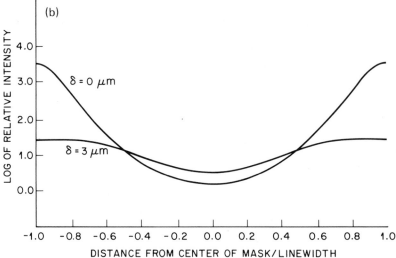

Fig. A-7. Variation of image intensity distribution with defocus. (a) Conventional image plot; (b) log-image plot. 1.0-μm space, 2.0-μm pitch, NA = 0.28, σ = 0.7, λ = 436 nm.

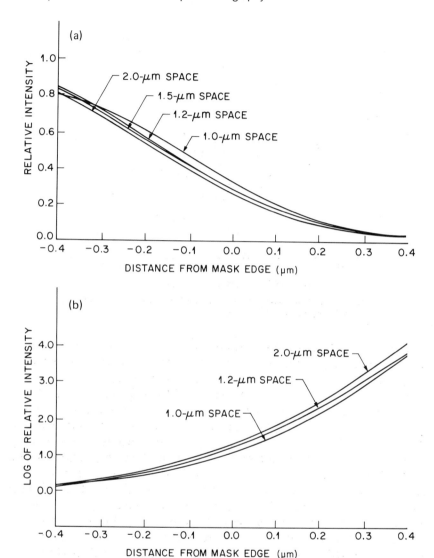

Fig. A-8. Comparison of image intensity distribution for different linewidths. (a) Conventional image plot; (b) log-image plot. 4.0-μm pitch, NA = 0.28, $\sigma = 0.7$, $\lambda = 436$ nm; no defocus.

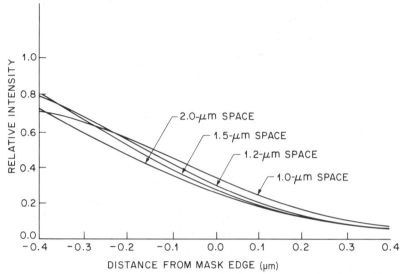

Fig. A-9. Comparison of image intensity distribution for different linewidths with 2-μm defocus. Pitch = 4.0 μm, NA = 0.28, $\sigma = 0.7$, $\lambda = 436$ nm.

sity. Shown in Fig. A-8 are portions of the image distribution near the mask edge (called the toe of the image) for different size mask features. It is very interesting that the 1.2- 1.5-, and 2.0-μm image distributions are almost identical in the toe region. Only the 1.0-μm feature deviates slightly from the rest. For the parameters shown, a 1.0-μm line corresponds to $k = 0.64$. Thus, the toe region of an image can be said to be independent of feature size until the feature approaches the theoretical resolution limit.

In Fig. A-9 the same features are defocused by 2 μm. Again note that the slope of the toe region is the same for each of the linewidths. The curves, however, are now spaced apart. This is due to the fact that the peak intensity of the feature is being decreased due to defocus, as was seen in Fig. A-7. The amount of this decrease is dependent on the feature size. Thus, for amounts of defocus that are not excessive, the major effect of defocus is the reduction of the peak intensity. Finally, Fig. A-10 shows how the slope of the toe region changes when the size of the feature is less than the theoretical resolution limit (which in this case is ~ 0.95 μm). The degree to which this effect is seen is a function of the coherence factor and the pitch. Other features that are in proximity to the feature in question will affect the image distribution of the feature (this is the so-called proximity effect of optical lithography).

The image intensity distributions shown assume perfect optical elements and thus represent best-case images for the given parameters. In reality, the

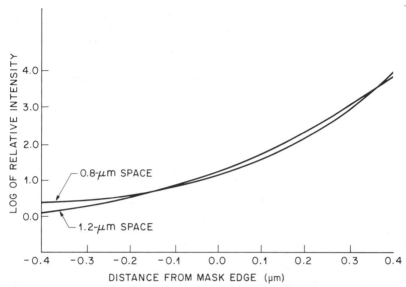

Fig. A-10. Comparison of image intensity distribution for different high-resolution line-widths (log-image plot). Pitch = 4.0 μm, NA = 0.28, λ = 436 nm; no defocus.

lenses used are not perfect and the actual image is degraded from the predicted image. For example, reflections and scattering from the lens surfaces result in a background intensity that may be 2–3% of the nominal intensity. Other lens imperfections, called aberrations, also degrade the resulting image. Thus, complicated lens arrangements and sophisticated lens optimization simulations are used to minimize the effects of these aberrations to give an image that is nearly perfect.

APPENDIX B: THEORY AND MATHEMATICS OF THE LUMPED PARAMETER MODEL

The lumped parameter model is based on a model for the development process, which in turn is based on the characteristic curve (also called the contrast curve) of a photoresist. Before proceeding, however, we shall set forth a few definitions. Let E be the exposure energy incident on the photoresist, $E(x, z)$ the actual energy at position (x, z) in the photoresist, and $I(x, z)$ the relative intensity within the photoresist. It is clear that

$$EI(x, z) = EI(x, z). \tag{B.1}$$

Furthermore, for projection printing it is a very good approximation to say that the intensity is separable, such that

$$I(x, z) = I(x)I(z). \tag{B.2}$$

We shall now define logarithmic versions of these quantities:

$$\epsilon(x, z) = \ln[E(x, z)], \quad \epsilon = \ln[E], \quad i(x, z) = \ln[I(x, z)],$$
$$i(x) = \ln[I(x)], \qquad i(z) = \ln[I(z)]. \tag{B.3}$$

Thus, Eqs. (B.1) and (B.2) become

$$\epsilon(x, z) = \epsilon + i(x) + i(z). \tag{B.4}$$

These logarithmic definitions will be useful in dealing with the characteristic curve of a photoresist, which uses $\log_{10} E$ as the abscissa.

Shown in Fig. 8 is a photoresist characteristic curve that uses the natural log instead of the conventional $\log_{10} E$. This curve relates thickness remaining after development to exposure energy. By examining the curve, one might expect that a reasonable fit to this curve can be obtained using an exponential function. In particular, the relative resist thickness remaining T_r can be modeled as

$$T_r = 1 - e^{\gamma(\epsilon - \epsilon_0)} \tag{B.5}$$

where ϵ_0 is the energy required to just clear the photoresist in the allotted development time. The constant γ is related to the contrast of the resist process, which we shall call γ_{10}, by

$$\gamma_{10} = 2.303\gamma. \tag{B.6}$$

The thickness remaining can be related to the development rate by

$$T_r = 1 - \frac{1}{D} \int_0^{t_{dev}} r(E) \, dt, \tag{B.7}$$

where D is the resist thickness, t_{dev} the development time, and r the development rate as a function of exposure energy. If the development rate is assumed constant through the resist (i.e., if resist absorption and standing waves are ignored), then Eq. (B.7) becomes

$$T_r = 1 - r(E)t_{dev}/D = 1 - r/r_0, \tag{B.8}$$

where r_0 is the development rate needed to just clear the resist in the allotted development time. Comparing Eqs. (5) and (8), one can see that

$$r = r_0 e^{\gamma(\epsilon - \epsilon_0)} \tag{B.9}$$

Although Eq. (B.9) was derived assuming no absorption or standing waves, the model can now be expanded to include these effects as simply a variation in the exposure energy:

$$r(x, z) = r_0 e^{\gamma(\epsilon(x,z) - \epsilon_0)} \qquad (B.10)$$

Equation (B.10) is a simple model relating development rate to exposure energy based on the common characteristic curve of a photoresist. It is also common to see curves that relate exposure energy and the size of a critical dimension (CD) feature. We shall now use Eq. (B.10) to derive an expression for CD versus exposure. This derivation is based on the assumption that the development process occurs in two steps: a vertical development to a depth z, followed by a lateral development to position x (measured from the center of the mask feature). This type of segmented development process is illustrated in Fig. 9.

A development ray, which traces out the path of development, starts at the point $(x_0, 0)$ and proceeds vertically until a depth z is reached such that

$$\epsilon(x_0, z + \Delta z) < \epsilon(x_0 + \Delta x, z). \qquad (B.11)$$

At this point the development will begin horizontally. The time needed to develop in both vertical and horizontal directions can be computed from Eq. (B.10). The development time per unit thickness is simply the reciprocal of the development rate,

$$\tau(x, z) = \tau_0 e^{-\gamma \epsilon(x,z)}, \qquad (B.12)$$

where

$$\tau_0 = e^{\gamma \epsilon_0}/r_0.$$

The time needed to develop to a depth z, t_z, is given by

$$t_z = \tau_0 e^{-\gamma \epsilon} e^{-\gamma i(x_0)} \int_0^z e^{-\gamma i(z')} \, dz', \qquad (B.13)$$

where Eq. (B.4) has been used. Similarly, the horizontal development time is

$$t_x = \tau_0 e^{-\gamma \epsilon} e^{-\gamma i(z)} \int_{x_0}^x e^{-\gamma i(x')} \, dx'. \qquad (B.14)$$

The total development time is fixed, however, at t_{dev} so that

$$t_{dev} = \tau_0 e^{-\gamma \epsilon} \left[e^{-\gamma i(x_0)} \int_0^z e^{-\gamma i(z')} \, dz' + e^{-\gamma i(z)} \int_{x_0}^x e^{-\gamma i(x')} \, dx' \right]. \qquad (B.15)$$

Equation (B.15) can be used to derive some interesting properties of the resist profile. For example, how would a small change in exposure energy,

$d\epsilon_0$, change the final position of the development ray? One can see that a change in exposure energy will not change the ratio of the energies given in Eq. (B.11). Thus, the depth z at which the ray turns to a horizontal direction will remain the same. The effect on the position x can be derived by differentiating Eq. (B.15), with the result

$$\left.\frac{dx}{d\epsilon}\right|_z = \frac{\gamma t_{dev}}{\tau(x,\,z)} = \gamma t_{dev}\, r(x,\,z). \tag{B.16}$$

Since the x coordinate is simply one-half the width of the feature (measured at a depth z), Eq. (B.16) defines the change in CD with exposure energy. Two forms of this equation will be given. The first can be obtained by taking the log of both sides to give

$$\ln(dx/d\epsilon) = \ln(\gamma t_{dev}) - \ln(\tau(x,\,z)). \tag{B.17}$$

Using the definition of $\tau(x,\,z)$ and Eq. (B.4), this expression becomes

$$i(x,\,z) = i(x) + i(z) = \frac{1}{\gamma}\ln\left(\frac{dx}{d\epsilon_0}\right) - \epsilon(x) + \frac{1}{\gamma}\ln\left(\frac{\tau_0}{\gamma t_{dev}}\right). \tag{B.18}$$

Equation (B.18) is the log-linear form of the lumped parameter model and relates the slope of a CD versus log-exposure curve to the image intensity.

A second form of the lumped parameter model can also be obtained from Eq. (B.16) by using Eq. (B.12) to give

$$d\epsilon/dx = e^{-\gamma(\epsilon(x,\,z)-\epsilon_0)}/\gamma D$$

$$= e^{\gamma(\epsilon(0)-\epsilon)}\, e^{\gamma(i(0)-i(x))}/\gamma D. \tag{B.19}$$

Invoking the definitions of the logarithmic quantities,

$$\frac{d\epsilon}{dx} = \frac{1}{\gamma D}\left[\frac{E(0)I(0)}{EI(x)}\right]^{\gamma}. \tag{B.20}$$

Also,

$$d\epsilon = dE/E \tag{B.21}$$

so that Eq. (B.20) becomes

$$\frac{dE}{dx} = \frac{E}{\gamma D}\left[\frac{E(0)I(0)}{EI(x)}\right]^{\gamma}. \tag{B.22}$$

This equation can now be integrated,

$$\int_{E(0)}^{E(x)} E^{\gamma-1}\, dE = \frac{1}{\gamma D}\, (E(0)I(0))^{\gamma} \int_0^x I(x)^{-\gamma}\, dx, \tag{B.23}$$

giving

$$\frac{E(x)}{E(0)} = \left[1 + \frac{1}{D} \int_0^x \left(\frac{I(x)}{I(0)}\right)^{-\gamma} dx\right]^{1/\gamma} \tag{B.24}$$

where $E(x)$ is the energy needed to expose a space of width $2x$ and $E(0)$ is the energy required to just clear the resist (i.e., to make a CD of 0). For larger lines, the peak intensity may occur at some point x_0 away from the center of the line. In such a case, $E(0)$ and $I(0)$ in Eq. (B.24) are replaced by $E(x_0)$ and $I(x_0)$, and the bounds on the integral are from x_0 to x.

Equation (B.24) can now be put in terms of log exposure:

$$\epsilon(x) = \epsilon(0) + \frac{1}{\gamma} \ln\left[1 + \frac{1}{D} \int_0^x \left(\frac{I(x)}{I(0)}\right)^{-\gamma} dx\right]. \tag{B.25}$$

Equation (B.25) is the integral form of the lumped parameter. Using this equation, one can generate a normalized CD versus exposure curve by knowing the image intensity $I(x)$, the resist thickness D, and the (base e) contrast γ.

REFERENCES

1. F. H. Dill, et al., IEEE Trans. Electron Devices **ED-22**, 445–452. (1975).
2. C. A. Mack, Opt. Microlithogr. IV, Proc., SPIE Vol. **538**, 207–220 (1985).
3. C. A. Mack, Adv. Resist Technol. III, Proc., SPIE Vol. **631**, 276–285 (1986).
4. C. A. Mack, Dispelling the myths about dyed photoresist (to be published).
5. C. A. Mack, J. Electrochem. Soc. **134**(1), 148–152 (1987).
6. C. A. Mack, Kodak Microelectron. Interface '85 (1985).
7. R. Hershel, Kodak Microelectron. Interface '78 pp. 62–67 (1978).
8. E. C. Kintner, Appl. Opt. **17**, 2747–2753 (1978).

Chapter 3

The Evolution of Electron-Beam Pattern Generators for Integrated Circuit Masks at AT&T Bell Laboratories

D. S. ALLES AND M. G. R. THOMSON

AT&T Bell Laboratories
Murray Hill, New Jersey 07974

I. INTRODUCTION

One of the most critical tools involved in the production of integrated circuits is a machine that employs an electron beam to pattern lithographic

57

masks. The machine is a computer-controlled servo system that directs an electron beam over a moving mask substrate to write patterns whose smallest dimensions are ~ 1 μm. At AT&T Bell Laboratories, where the development of electron-beam mask making began in the early 1970s, such electron-beam exposure systems are denoted by the acronym EBES.

In this chapter, we shall describe not only what mask patterning entails and how it is done but also the AT&T Bell Laboratories EBES4, the newest of our electron lithographic systems. We shall begin by describing the product that EBES generates—the integrated circuit mask—and the requirements placed on the mask by the mask users. More detailed descriptions have previously been given by Herriott and Brewer [1] and by Ballantyne [2].

II. INTEGRATED CIRCUIT MASKS

In the fabrication of an integrated circuit, a semiconductor wafer (see Fig. 1) is coated with a photosensitive layer (photoresist) and the latter exposed by the pattern of light passing through the *optical mask* (Fig. 2). The mask is usually imaged onto the wafer by means of a projection camera. Development of the exposed resist removes the exposed or unexposed resist (depending on whether the resist is positive or negative) and exposes the underlying wafer in a pattern through which etching, ion

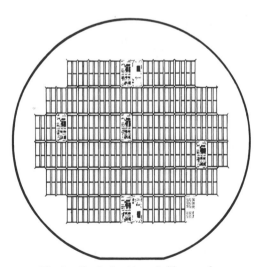

Fig. 1. Typical patterned silicon wafer.

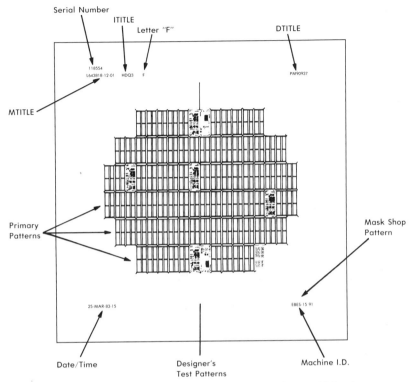

Fig. 2. Chromium-covered glass mask showing some of the additional patterns and human-readable information.

implantation, metallization, etc., can be carried out. Depending on the type of integrated circuit being fabricated (NMOS, CMOS, bipolar, etc.), there can be between 6 and 25 lithographic steps, each requiring a separate mask with patterns defining a different set of circuit features. The optical mask is usually a glass plate on which a thin layer of chromium has been patterned, as shown in Fig. 2. Each of the identical rectangular areas contains the pattern for one level of an integrated circuit.

The need for multiple levels of lithography per wafer gives rise to two general mask requirements: (1) Each mask in the set must be accurately made so that the patterns defined on the semiconductor wafer at one lithographic step will precisely align with the patterns printed by all subsequent masks in the set, and (2) there must be very few defects on the masks, since even a small defect may destroy the circuit on which it occurs. With six or more lithographic steps, a few randomly placed defects per mask will result in a significant reduction in the final yield of good circuits per wafer. Thus, mask makers must continually strive both to improve the accuracy and to reduce the defect density on their masks. Since 1970, the improvements in both areas have been dramatic, with overall accuracies improving from several micrometers to 0.1 μm and defect densities improving from more than 10 defects to fewer than 0.2 defect per square centimeter. However, the continuing reduction in integrated circuit linewidths and increase in the dimensions of each complete circuit (the chip size) maintain the pressure on mask makers to provide further improvements in the future.

A. Mask Types

Masks can be classified as full field or reticle. A full-field mask is printed with approximately unity magnification and hence is also referred to as a 1 $\times$ mask. A reticle is usually demagnified by a factor of 5 or 10 and is thus called a 5 $\times$ or 10 $\times$ mask.

As the name implies, *full-field* masks are those that, in a single exposure, transfer to the resist-covered wafer all the patterns corresponding to the array of chips on the wafer. The mask may be placed very close to the wafer and illuminated or it may be placed in a one-to-one projection camera and its image projected with unity magnification onto the wafer. Chips are usually arrayed on the wafer in a circular area, currently as large as 125 mm in diameter. The mask printed one-to-one must also be this large, and this places an additional restriction on mask fabrication: The thermal expansion and therefore the temperature of the mask must be strictly controlled to ensure the exact registration of one lithographic level with the next. A temperature change of 1 °C will result in a more than 1.2-μm error at the edge of a 125-mm glass mask. To help, ordinary glass for masks is being replaced by low-expansion (Pyrex-like) glass and by fused silica. The last, in addition, better transmits the ultraviolet light used to illuminate the mask and expose the photoresist.

Reticle is the name given to masks with patterns corresponding to only one or perhaps a few of the chips in the entire wafer array. Reticle patterns are imaged onto the wafer with a *step-and-repeat* camera whose image field may be only 1 cm². The camera produces the desired full-field array of chip patterns by repeatedly exposing the resist-covered wafer to the reticle image and precisely stepping the wafer to a new position before each succeeding exposure.

In the case of a single-chip reticle, the reticle pattern may be the same size as the chip or a 5–10 $\times$ enlargement. Larger dimensions, of course, make it easier to write the mask features. The features and any errors in the features are demagnified by the corresponding optical reduction built into the step-and-repeat camera.

Advantageous as this may be, the gain is bought at a cost. Since exposure of the reticle image is repeated many times over the wafer, even a single defect on the reticle will appear at each site on the wafer. If the defect is a fatal one and the reticle contains only one chip pattern, the resulting integrated circuit device yield from wafers so exposed will be zero. And, of course, the larger the reticle, the larger the area that must be defect free. For example, a 5× enlargement reticle must be defect free over an area 25 times that of the chip area (but much larger defects can be tolerated before they become fatal).

In an emerging technology, masks are being printed on wafers using x radiation. The mask substrates required for use at these wavelengths differ from conventional mask substrates in that they must be transparent to x rays rather than to visible or ultraviolet light (Fig. 3). They are commonly thin membranes of material such as silicon or silicon carbide. The features that are intended to be opaque to the x radiation are patterned in gold deposited on the membrane. X-ray masks are always exposed with a magnification very close to unity. They are placed close to but not in contact with the wafer during exposure. (X-ray projection cameras are not available.) Diffraction, a major cause of resolution loss when ultraviolet light is used to illuminate optical masks, is a less critical limitation at the short wavelengths of the x radiation but is significant when feature sizes approach 0.25 μm. The use of x rays offers the promise of much better feature resolution, but control of mask distortions over a field of only a few centimeters diameter is difficult.

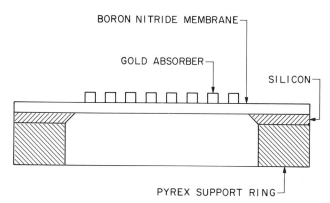

Fig. 3.   Cross section of an x-ray mask.

B. Mask Requirements

Mask specifications imposed by the designers of integrated circuit chips are stringent:

(1)   Defects greater than 1 μm in dimension must have a density of 0.2/cm^2 or less.

(2)   Linewidths must be correct to within 0.1 μm.

(3)   Features must be positioned correctly to within 0.125 μm over the entire mask surface.

Without this level of perfection, the mask is useless. It will result in few if any good chips. In addition, the designers need several pattern options to maximize the chip yield and performance and to simplify the wafer fabrication process. In many cases these options cause the mask writer to alter the way the pattern is represented on the mask without the need to alter the original chip design information explicitly, and it must lie within the capability of the mask pattern generator to provide these options:

(1)   In addition to the primary chip pattern, masks may include many other patterns (Fig. 2) for alignment, in-process testing, and identification.

(2)   Chip patterns on the mask may have clear features on an opaque background or opaque features on a clear background. This may be specified on a chip-by-chip basis or over the entire area of the mask.

(3)   An integrated circuit is usually designed with the edges of the features lying on a grid with a certain address size. It may be designed with a wide range of these address sizes, and the pattern generator must adjust its address size to match that of the chip.

(4)   The entire mask pattern may be mirrored (about the mask's centerline) or each of the chip patterns may be mirrored separately (about their centerlines) or both.

(5)   The chip patterns may be magnified or demagnified relative to their original design information to improve manufacturability or performance or both. This option does not alter the chip's location on the mask.

(6)   The entire mask may be magnified or demagnified to correct for temperature mismatches between the mask writing environment and the mask usage environment or to correct for wafer size changes resulting from the application of stressed films during the integrated circuit manufacturing process.

(7)   Human-readable information may be requested, including the mask number, the pattern generator identification, the inspection criteria, the date written, and the mask serial number.

(8)   Special patterns may be included to monitor the performance of

the pattern generator and of the mask shop's resist-development and chrome-etching processes.

(9)   The type of glass requested may be white crown, low expansion, or quartz, and the type of chrome may be normal or low reflectivity. The type of resist may be positive or negative.

(10)   The exposure level may be modified to cause a slight increase or decrease in the size of every feature, without altering the location or size of the chips.

Most masks require several of these options.

C. Mask Fabrication

The mask fabrication process begins with the mask designer bringing to the mask shop magnetic tapes containing the geometric descriptions of patterns on each chip and a description of the mask array, indicating not only the location of each chip but also the options to be used when writing and processing the mask. In our mask shops, this information is entered into a computer system, known as the mask-shop information and management system (MIMS), which

(1)   checks the input information for consistency and data errors,

(2)   preprocesses the data, if necessary, to make it compatible with the pattern generator,

(3)   schedules the mask to be written on a specific pattern generator and sends the chip pattern and array data to that pattern generator, and

(4)   schedules the exposed mask to be developed, etched, inspected, and (if necessary) repaired.

At every step of the process (writing, developing, etching, repair, and inspecting) the information system collects information that is used to monitor the pattern generator performance, and this information is fed back to the shop operation and pattern generator maintenance personnel.

D. Generating Masks with an Electron Beam

In the past masks were made by mechanized drafting machines that drew the pattern of a single chip at 250 times the final size. These patterns were later reduced to make $10\times$ reticles that were further reduced and arrayed by the step-and-repeat camera to make the final mask. Later, optical pattern generators became popular. They were step-and-repeat-like machines with variable-sized rectangular apertures whose images were reduced and projected onto a photosensitive plate. By proper exposure of sequences of these images, they could compose a 10×-magnified reticle of the chip pattern. Still later, in the mid-1970s, AT&T Bell Laboratories developed the electron-beam pattern generators EBES I and EBES II that were capable of writing one-to-one masks directly. Direct writing of masks eliminated most of the steps in the mask-making process and greatly reduced the time to make complicated masks.

The early electron-beam systems employed a small-field scan technique:

(1)   The beam was scanned across a narrow (256-μm-high) stripe (see Fig. 4), and it was turned on and off to "paint in" the desired pattern.

(2)   Between scans, the beam's position was advanced one address along the stripe.

(3)   An interferometer and electron deflection system corrected the beam's position for small stage-position errors, thereby allowing the writing to proceed even though the stage was not stationary.

(4)   The pattern data were stored in a memory in which each bit corresponded to an address in the narrow stripe. These data were read in synchronism with the beam's scan and the beam was turned on or off as required.

The speed and simplicity of this design made these systems reliable and removed the uncertainty about mask delivery times.

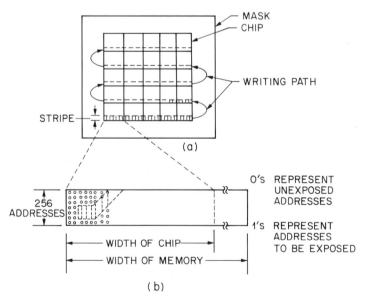

Fig. 4.   Raster scan scheme used in the early EBES systems.

The design of EBES II was later licensed to two companies: ETEC (later purchased by Perkin Elmer) and Extrion (later purchased by Varian). Using AT&T Bell Laboratories' technology and drawings, these two companies have improved, built, and delivered more than 75 EBES-like mask pattern generators to mask-making companies and semiconductor manufacturers throughout the world, and they are now the pattern generators used to write most of today's most demanding masks.

III. A NEW SYSTEM: EBES4

The commercial EBES-like electron-beam pattern generators have displaced all of the optical pattern generation systems in AT&T Technologies Systems (formerly Western Electric) mask shops. They routinely operate 24 hr/day, 6 or 7 days/week and have better than 90% uptime. Their performance and acceptance have surpassed both our original goals and our best expectations. The long-term need for more of these pattern generators to meet our mask-making needs was recognized in the late 1970s when the EBES design was licensed to ETEC and Extrion. We at AT&T Bell Laboratories recognized that ETEC and Extrion would pursue an aggressive program of evolutionary development and that we could be assured of a supply of compatible (although probably better and faster) EBES-like systems from these vendors in the near future. Therefore, we decided to look further into the future and to design an electron-beam exposure system to meet new specifications appropriate for the most advanced integrated circuits.

We intended to design the new system from the ground up, basing the design on new, more stringent specifications and paying heed to all the lessons that had been learned during the design, fabrication, and final development of the previous EBES systems.

A. The New Specifications

The specifications for EBES4 were developed in 1977 as a result of discussions with the mask shops and integrated circuit designers. It seemed clear that a finer address structure would be required, that the ever expanding need for masks would require even higher throughputs, and that larger mask and wafer substrates would have to be accommodated. These needs were recognized in the following goals:

(1) Nominal address size: 0.125 μm (but needed to be variable from 0.06 to 0.125 μm)

(2)   Throughput: two 100-mm-array masks per hour

(3)   Accuracy: 0.1 μm from all causes over a 100-mm array

(4)   Maximum array size: 150 mm

(5)   Maximum substrate size: 200 mm

(6)   Primary product: one-to-one optical masks (although direct wafer writing and magnified reticles could also be accommodated)

(7)   Input data from the designers: compatible with previous EBES systems

A system meeting these specifications represented a significant improvement over previous EBES systems. To compare, the latter offered a variable address size ranging between 0.25 and 0.5 μm, and at a 0.25-μm address size the time to write a 100-mm-array mask usually exceeded 2 hr. Errors were sometimes as large as 0.25 μm and a 100-mm array could just be accommodated. Thus, EBES4 was to write faster, finer, more accurately, and over a larger area than any previous electron-beam pattern generator.

Although some things have changed during the past few years these

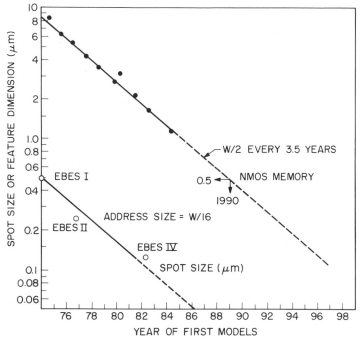

Fig. 5.   Feature size used in production integrated circuits and address size in the best available mask maker.

specifications still appear to be suitable. (Figure 5 shows both the minimum linewidth used in production memory circuits and the pattern generator address size versus time.) However, a significant change has been the emergence of step-and-repeat cameras for exposing the most critical integrated circuit wafer patterns and the resulting need for 5×-magnified reticles rather than one-to-one masks.

Although we cannot predict the future, EBES4 appears to be well positioned to meet the needs of the device designers and process lines because it patterns both masks and reticles with high precision and speed, and it opens the door to direct electron-beam exposure experiments and short-run production of fine-line devices.

B. Alternate Writing Approaches

To increase the rate of area coverage by an electron-beam pattern generator, several approaches can be taken:

(1) Increase the resist sensitivity. With the same electron column and source but with increased deflection rates, less time is required to expose each address and the exposed area per unit time increases.

(2) Increase the beam size at constant current density. With the same resist a larger area is exposed in the same time. This is particularly attractive if the spot size or shape (or both) can be changed, thereby allowing the spot size and shape to match the features that are being exposed.

(3) Increase the source brightness and deflection rate. With the same electron resist, less time is required to expose an address and the exposed area per unit time increases.

Before settling on the EBES4 writing philosophy, we investigated each of these approaches. Each approach could meet the new specifications, but after some investigation and experimentation, the first two were deemed to have more drawbacks than the third (selected) approach.

Increasing the sensitivity of the resist initially appears to be an attractive approach because it allows any system to be run more quickly (if the electronics and deflection systems can support the greater data rate). Unfortunately, it is the statistics of the exposure process that limits the usable resist sensitivity. If the resist becomes more sensitive, the number of electrons that will be deposited in an address area decreases. However, the arrival of electrons at the address area on the resist is a random process, and unless the number of electrons in an address is statistically large, the level of exposure at that address is uncertain. Consequently, randomly placed addresses may be under- or overexposed, thereby diminishing the quality of the lithography.

Increasing the system throughput by increasing the area of the writing beam is attractive, provided that the exposing beam's area remains smaller than the minimum feature. However, this approach is not without drawbacks. As the spot size increases, the total beam current increases, and eventually the density of electrons in the beam becomes so great that the electrons in the beam begin to influence one another's trajectories. At this point the electron optical performance of the deflection and focusing optics is degraded, with a similar reduction in the sharpness of the exposed pattern.

Our first attempt to use a larger-area beam in order to increase throughput was an extension of the EBES raster scan scheme described earlier [3]. This employed two specially shaped electron-illuminated apertures (Fig. 6a), one of which was imaged onto the other, and the composite was, in turn, imaged onto the writing surface (Fig. 6b). By rapidly deflecting the position of the first aperture image relative to the second while the composite writing image was scanned, one in effect created four beams whose on–off condition could be controlled independently. This would have allowed raster-scanning EBES machines to write four times as fast as the original single-beam system. This increase in the area exposed when four beams were used did not increase the throughput sufficiently even when this was combined with a bright lanthanum hexaboride cathode, and after building a demonstration column, we ceased work on this approach.

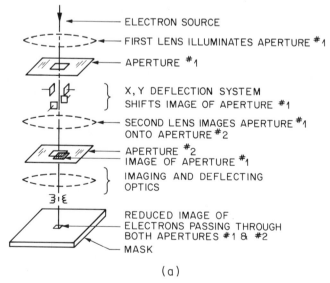

(a)

Fig. 6. (a) Schematic of a variable-shape spot column; (b) spot-forming aperture shapes used to produce four parallel scanned beams. (From Thomson *et al.* [3].)

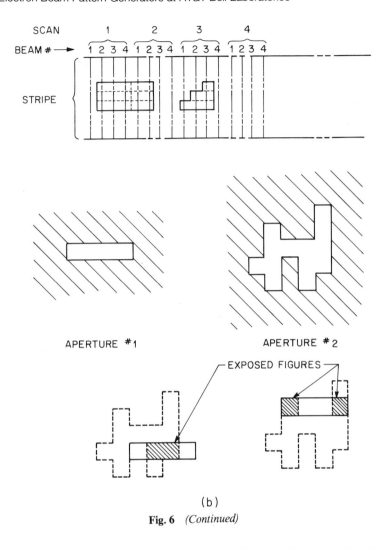

(b)

Fig. 6 *(Continued)*

An extension of this scheme was to expose figures with a variably shaped beam whose area was larger than a single address and to build patterns by filling them in with appropriate beam shapes. Figure 7 illustrates one possible set of the aperture shapes. These apertures were used in an electron column similar to that shown in Fig. 6 except that the resulting figures were not scanned. The shape and position were set by the deflection system, and the beam was turned on (unblanked) for a time long enough to expose that figure in the resist. The deflection system then moved the beam to the position of the next figure, adjusted the beam shape, and exposed the

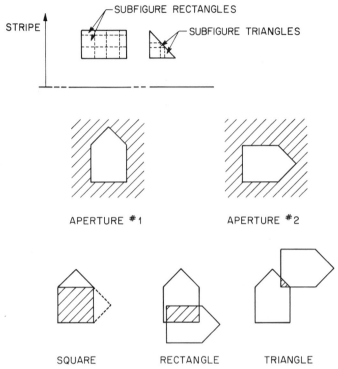

Fig. 7. Examples of spot-forming aperture shapes used to produce rectangular and 45° triangular spots.

new figure. The advantage of this approach was that the beam did not have to scan over the entire area of the mask, as it did in the raster scan approach of previous EBES, but needed only to address the areas that were actually exposed. The latter is often referred to as vector scan writing rather than raster scan.

Although this approach looked promising for small shapes, the larger beam current accompanying large shapes increased beam broadening due to the electron–electron interactions, and the feature edge resolution was unacceptably degraded. In addition, the increased space charge within the beam caused the focal point to be displaced below the substrate surface. This caused an additional degradation in edge resolution for all feature areas except that for which the system was aligned. We deemed this to be a major impediment to finer pattern lithography, and this approach was not pursued further. Other manufacturers, however, have adopted this approach, and several system have been produced or are under active development [4–6].

Turning our attention toward the third approach, increasing the source brightness, we found that the state of the art of field-emission electron sources had advanced to the point where they might be considered for electron-beam lithography. Thermal field-emission sources (TFE) not only could produce the necessary current for a high-throughput lithographic system using robust resists but also would be bright enough to allow extremely short address exposure times. A similar route has been taken by Eidson *et al.* [7] at Hewlett-Packard Laboratories, although their application was writing directly on silicon wafers with a larger spot size (0.25 – 0.5 μm).

Both the shaped-beam and the TFE systems were evaluated, and the decision to use the small-beam TFE approach rather than the variably shaped beam approach was based on several factors:

(1) The TFE column could be simple and short, thereby minimizing the electron–electron effects.

(2) As the feature sizes decrease, the throughput of a system using the TFE column degrades more slowly than that of one using a shaped-beam column.

(3) The throughput of the TFE approach can be made less sensitive to arbitrary slant-line geometries than the shaped-beam approach.

(4) The number of apertures illuminated by the electron beam in the shaped-beam column and the fact that their electron illumination varies with time led to concerns about electrostatic charging that would result in beam-shape and exposure instability.

For these reasons we chose the TFE approach, and the remainder of this chapter describes the machine that was designed and constructed.

IV. OVERALL SYSTEM DESCRIPTION

EBES4 [8] uses a small-diameter, high-brightness, Gaussian-shaped electron beam to expose the required patterns. The exposure is accomplished by raster scanning the beam (by stepping it from point to point) over the area of the pattern figure to be exposed and then moving the beam to the next figure (as opposed to rastering the beam's position over the entire area and turning it on only when the beam position is over a figure to be exposed). To expose the entire area of the substrate the patterns are broken into subfields each of which is exposed sequentially as the continuously moving stage brings the area to be exposed within the deflection range of the electron beam.

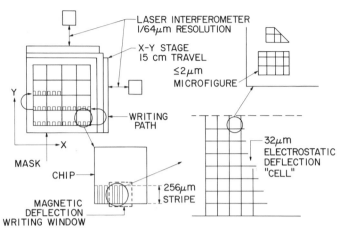

Fig. 8. Stage movement together with three deflection systems are used to fill in the pattern. (From Alles *et al.* [8].)

Figure 8 illustrates the hierarchy of deflection systems, and Fig. 9 illustrates how the 4-μm deflection system is used to expose an integrated circuit mask. The moving stage supports the mask substrate, and it has an overall travel of 15 cm in the X direction and 20 cm in the Y direction. The stage is used to bring the area to be exposed within the $\pm$ 128-μm "writing window," that is, the deflection range of the electron-beam column's magnetic deflection system (known as the 256-μm deflection system). This deflection system and the stage drive system operate in concert to stabilize the electron beam's position on the substrate even though the substrate (clamped to the stage) is moving with respect to the column. During writing, the computer requests that specific locations on the mask substrate be brought under the electron column's centerline. These locations are referred to as the wish you were here (WYWH) locations. The stage drive system moves the stage to minimize the error between the actual stage location (as defined by the intersection of the electron column centerline with the substrate) and the WYWH. It also calculates the actual stage-position error and directs the 256-μm deflection system to correct for the remaining error. If the error is too great to be corrected by this deflection system, writing is halted until the stage drive system once again brings the area to be written within the $\pm$ 128-μm writing window.

In addition to the moving stage and 256-μm magnetic deflection systems, there are two short-range electrostatic deflection systems, having nominal ranges of 32 and 4 μm, respectively. The 32-μm deflection system is used to position the beam (its deflection adds to the 256-μm deflection) at the origin of every figure, and the 4-μm deflection system moves the

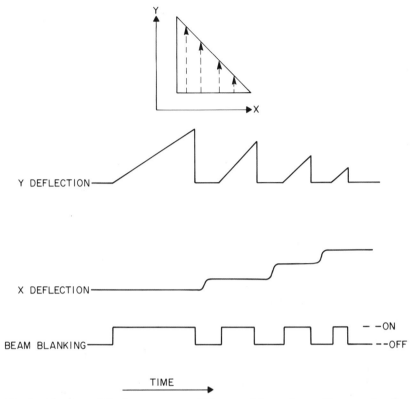

Fig. 9. The 4-μm deflection system can fill in several figure shapes. The procedure for a 45° triangle is illustrated.

beam during the exposure process to "paint in" the desired figure shape. One additional deflection system, the beam blanker, turns the beam on and off but is not used to change the *position* of the beam.

Although each of the deflection systems has a different range (256, 32, and 4 μm) and each has a different settling time (10 μsec, 10 nsec, and <1 nsec, respectively), they all have an absolute accuracy of $\sim\frac{1}{64}$ μm. The position of the stage is also known with the same precision. This hierarchy allows a large area to be patterned with great precision and in a minimum amount of time.

Because the pattern on *each* chip accumulates (as each new stripe is written) throughout the exposure of the *entire* mask, the long-term stability of the system is critical. Any drift during the writing of the mask will create a distortion in every pattern. For this reason, EBES4 has been designed with careful attention to thermal stability, magnetic shielding, and electronic stability. In addition, a fiducial mark on the stage is periodically scanned to measure, and then correct for, small drifts between the stage and the electron-beam column.

The discussion of EBES's component subsystems is divided as follows: data handling, writing strategy, electron-beam column, mechanical systems, control system, and software.

A. Data Handling

The structure of the pattern data required at writing time by EBES4 is different from that for EBES I and II. Overlapping figures must be removed before exposure, and different data are required depending on the "tone" of the pattern (clear features on an opaque background or opaque features on a clear background). In previous generations of EBES a bitmap of the pattern was created that was one stripe high (i.e., 256 μm at 0.5 μm address size) and as wide as the pattern. Every pixel in the stripe was represented by one bit in the memory, and the data representing the figures were loaded into the memory by OR-ing a "1" into the corresponding memory location. This automatically removed the overlapping figures, and either tone could be exposed by turning the beam on when the bit was either a "1" or a "0".

On the other hand, EBES4 exposes figures rather than single addresses in a raster fashion, so overlapping figures (see Fig. 10) will create doubly exposed areas (producing unacceptable variations in the widths of the developed lines). Moreover, a different set of figures is required to reverse the tone. Since all of these machines, new and old, must accept pattern data in the standard AT&T EBES format, EBES4 must preprocess its pattern data to remove overlaps and adjust the tone.

Our approach to pattern preprocessing has followed two different routes: a purely software solution and a software solution augmented by special-purpose hardware. Our initial attempt was to use software, but the time required was excessive, particularly for very complex patterns. Therefore, we designed a hardware accelerator peripheral (the pattern processing unit, or PPU) for the VAX™ 11/780 that assisted in the overlap removal and tone reversal calculations. The result was a hardware–software system that could preprocess patterns at a rate that was almost independent of pattern complexity but directly dependent on the overall pattern (chip) size. The pattern data for the metal level of a 1-Mbit DRAM can be preprocessed in 36 min elapsed time (20 min CPU time) on an unloaded VAX 11/780. When the same pattern is preprocessed for writing as a 5× reticle (so that the area of the chip was increased 25 times), the preprocessing time required is 254 min of elapsed time and 92 min of CPU time.

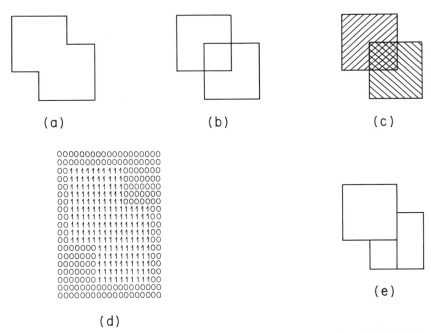

Fig. 10. Overlapping figures are removed in different ways on EBES I and EBES4. The desired final figure shape is shown in (a). Shown in (b) is the way a designer might describe the figure as two overlapping simpler shapes that would result in a figure with a double-exposed region (c). In EBES I, the two shapes are "OR"-ed into the buffer memory (d), so that the correct final figure is written. In EBES4, the pattern data must be preprocessed to break the figure into nonoverlapping simple components, as in (e).

The output from the PPU consists of descriptions of 32-μm cells that eventually "tile" the area of the chip. The output is written onto a disk that can be read both by the computer used for the preprocessing and by the EBES4 control computer.

B. Writing Strategy

Referring to Fig. 8, the writing process is as follows:

(1) Using pattern data that have been prepared previously, the WYWH location of the first area to be patterned is given to the stage drive system, and it drives the stage until this location is close to the electron column's centerline. As the error becomes less than ± 128 μm, the stage control uses the 256-μm magnetic deflection system to correct for the remaining stage-position error.

(2) When the stage error is within the range of the magnetic deflection

system, writing begins. The magnetic deflection system directs the beam to the center of the first 32-μm (nominal) subfield known as a cell. The 32-μm deflection system then deflects the beam to the origin of the first figure and the 4-μm deflection system paints in the figure.

(3) The 4-μm deflection system exposes figures by rastering the beam point by point over the figure's area. Its repertoire of figures contains rectangles, four orientations of 45° right triangles, and two 45° parallelograms. In all cases, the figures can be from 1 address ($\frac{1}{8}$ μm) to 16 addresses (2 μm) on a side.

(4) The 32-μm deflection system deflects the beam sequentially to each of the figures within the cell, and the 4-μm system exposes the figures.

(5) The 256-μm deflection system sequentially deflects the beam to every cell across the height of the 256-μm stripe. After all of the cells in this group have been patterned, the WYWH is moved to address the next vertical row of cells. This process occurs so rapidly that the stage can move continuously.

(6) Each of the above steps is repeated until the entire stripe along the first row of chips is exposed. The first stripes of the second and succeeding rows of chips are patterned in a serpentine fashion until the first stripe has been exposed on all chips.

(7) The process is repeated for successive stripes until the entire mask has been exposed.

C. Electron-Beam Column

The electron column [9] (see the cutaway view in Fig. 11) provides the source of electrons, and focuses, blanks (turns on and off), and deflects these electrons over the small writing window on the mask. The requirements from the EBES4 electron column are as follows:

(1) 250 nA into a 0.125-μm-diameter spot at 20 keV
(2) Less than 1% spot current variations during the writing of a mask
(3) Greater than 2000 hr source lifetime
(4) Less than 0.5-nsec beam blank and unblank times
(5) Deflection systems:
  (a) Long-range 256-μm magnetic deflection systems—settling time < 10 μsec to an accuracy of $\frac{1}{64}$ μm (1 part in 16K)
  (b) Intermediate-range 32-μm electrostatic deflection system— settling time < 100 nsec to an accuracy of $\frac{1}{64}$ μm (1 part in 2K)
  (c) Short-range 4-μm electrostatic deflection system—settling time of < 1 nsec to an accuracy of $\frac{1}{64}$ μm (1 part in 256)
(6) Dynamic focus range ± 30 μm—settling time < 10 μsec

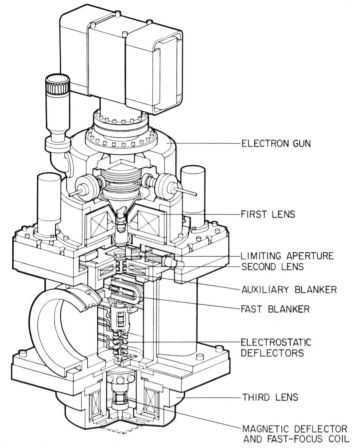

Fig. 11. Cutaway view of the electron-beam column. (From Thomson *et al.* [9].)

(7) Random position errors due to all causes (magnetic, mechanical distortion, vibration, thermal distortion) $\leq 0.1 \ \mu m$

Figure 12 is an electron optical schematic diagram of the electron column showing the locations of the magnetic lenses, deflection and blanking systems, and an electron ray diagram that traces typical electron trajectories from the source to the mask. The TFE electron source is immersed in the first lens. This lens collects the broad angular distribution of electrons from the cathode's tip and directs them through the beam-defining (or -limiting) aperture. The next (second) lens focuses the electrons to a diameter of $\sim 1.0 \ \mu m$ at the knife edge in the beam-blanking system. Below the fast blanker, three sets of X and Y electrostatic deflection plates make

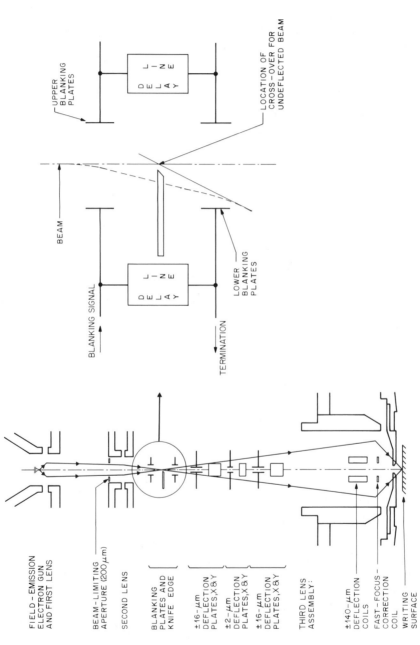

Fig. 12. Diagram of the optical components in the column. The centering coils and stigmators have been omitted. The diagram is to scale with two exceptions: The separations between all the electrostatic deflection and blanking plates have been exaggerated by a factor of 2, and the radial extent of the electron trajectory by a factor of 100 (both with respect to all the other dimensions in the diagram). (From Thomson *et al.* [9].)

up the 4- and 32-μm deflection systems. The final (third) lens demagnifies the 1.0-μm spot size at the beam blanker by a factor of 8 and focuses the beam at the writing surface. The 256-μm magnetic deflection coils are housed within the final lens. The field produced by these coils and by the final lens make the beam deflection telecentric; that is, the chief electron ray is always perpendicular to the writing plane. Therefore, a telecentric deflection system minimizes those beam-position errors that occur in non-telecentric systems when the writing surface is not in the focal plane.

Each of the major electron column subsystems will be described in the following sections.

1. Electron Source and Gun

The electron gun is the mechanical structure that supports the TFE source and provides the electrical, thermal, and vacuum environment for the source. The EBES4 gun is removable so that a spent source can be replaced with a new pretested source in a moderately short time (1 – 2 days). (This time is acceptable because the source life exceeds 5000 hr.)

The electron source, or cathode, is a pointed single crystal of $\langle 100 \rangle$ tungsten welded to a tungsten "hairpin"-shaped heater. Along the shank of the tungsten crystal, a source of elemental zirconium is deposited as a powder, fused to the crystal, and processed in low-pressure oxygen [10,11]. This zirconium diffuses to the 1.0-μm-radius tip of the cathode and reduces the work function of the tungsten, thereby enhancing its brightness.

The cathode, when mounted in the gun, is surrounded by electrodes (Fig. 13). The cathode's operating temperature is 1750 – 1800 K, and it operates at -20 kV with respect to ground. The shield electrode (-20.5 kV) minimizes the number of thermally generated electrons that will enter the main beam. The first anode, or extraction electrode (approx. -8 kV), "pulls" electrons from the tip of the tungsten needle by electrostatic force. The electrons that pass through the hole in the first anode are accelerated to the full beam potential of 20 kV by the grounded second anode.

TFE cathodes have been of academic interest for several years, but because of an output current instability, they have been unsatisfactory for electron-beam lithography applications. Liu [12] at AT&T Bell Laboratories has studied the problem and found a stable operating region where the thermal (surface energy) forces, which tend to blunt the tip and increase the radius at the end, are just balanced by the electrostatic forces that tend to pull the tip out into a needle point. This research has resulted in both long- and short-term emission current stability of better than 1% over the life of the source.

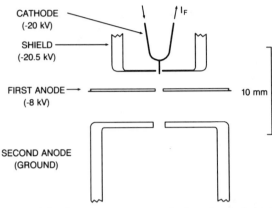

Fig. 13. Diagram of the zirconium–tungsten cathode and gun electrodes. (From Thomson *et al.* [9].)

Our experience indicates that several thousand hours of operation can be expected before the source of zirconium is depleted, and the emission current abruptly decreases to zero.

The accelerator within the gun was chosen to have little focusing action because of the inferior focusing qualities of electrostatic electron lenses. Instead, the beam is collimated by a large-bore magnetic lens within which the entire cathode and accelerator structure is immersed.

2. Electron Lenses

The magnetic structure of the first and second lenses is entirely iron, but the third (final) lens is a composite of iron and ferrite. The ferrite is necessary because this lens houses the deflection coils and the fields from these coils not only interact with the beam but also with the surrounding magnetic material. If the magnetic material is a conductor, the changing flux will produce eddy currents, and these currents will also produce a beam deflection that will persist over a period of many milliseconds. Because the magnetic deflection system must displace the electron beam to a new and stable location within microseconds, the interior of the final lens was constructed of a nonconducting manganese–zinc ferrite.

A small fast-focus coil is also immersed in the final lens to allow rapid refocusing over a range of $\pm 32\ \mu$m while writing. This is necessary because the resist-coated surface of the mask is not always in the nominal focal plane of the final lens because of mechanical uncertainties in the stage and the cassette that holds the mask. The current in this coil is varied as the stage is moving to maintain the beam in focus on the surface of the substrate. The additional "cyclotron" rotation produced by this coil is less than $0.1\ \mu$m over the 256-μm field; however, the deflection control system corrects for the resulting error in both the deflection gain and rotation.

Both the first and third lenses are moderately strong, and as a result, their coils dissipate enough power to create temperature gradients in the column. For this reason, both lenses are surrounded by water-cooling coils, to remove their heat and to minimize the temperature gradients throughout the machine.

3. Magnetic Deflection System

The large (256-μm) deflection system is magnetic rather than electrostatic because magnetic deflection systems can be designed to have smaller aberrations (distortion, defocus, coma, etc.) than electrostatic deflection systems of similar strength. The magnetic deflection field is produced by two carefully designed coil pairs immersed in the final lens (see Fig. 12). Their center of deflection is coincident with the back focal plane of the final lens, and as a result, the deflection system is telecentric. As with the lenses, the mechanical precision of these coils is crucial to their performance. Each coil consists of only a few turns, thereby minimizing the coil's inductance and allowing a fast current amplifier to alter the beam deflection and to settle to 1 part in 16,000 within a few microseconds.

4. Electrostatic Deflection Systems

The shorter settling time requirements of the 32-μm (100 nsec) and 4-μm ($<$ 1 nsec) deflection systems require an electrostatic rather than a magnetic deflection system. Because the deflection ranges are substantially smaller than that of the 256-μm deflector, the large inherent aberrations of an electrostatic system are tolerable.

Two sets of X and Y deflection plates (Fig. 12) are used in the 32-μm deflector. These are driven so that their effective center of rotation coincides with the back focal plane of the final lens. Therefore, both the 32- and 256-μm deflection systems are telecentric.

The 4-μm deflection system uses only one set of X and Y plates. Because these must settle in a time that is short compared with the transit time of the signal from the driver to the deflection plates, they are driven with a doubly terminated 50-Ω transmission line with a 50-Ω resistor immediately adjacent to the plates.

5. Beam Blanker

The beam must be turned on and off in a time that is short compared with one address time (2 nsec), and the beam position, on the mask substrate, must not move during blanking or unblanking. This is done by putting the electrostatic beam blanker's center of deflection at an intermediate image (Fig. 12). As a result, the beam moves off the column center-

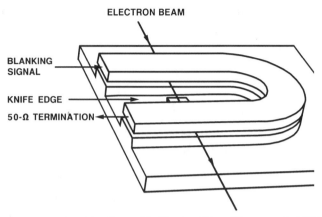

Fig. 14. Blanker delay line and knife edge. (From Thomson *et al.* [9].)

line as the beam-blanking deflection field increases, but because the center of deflection is at the beam crossover the apparent location of the crossover does not move and the action of the final lens maintains the beam's location fixed on the stage. As the beam-blanking deflection field increases, the beam eventually strikes the blanking knife edge, and the beam current is cut off. The upper plates deflect the electrons onto the blanking knife edge, and the lower plates deflect the beam an additional amount so that the center of deflection (while looking up the column) appears to be at the knife edge. This clever approach was first used on the Hewlett-Packard electron-beam pattern generator as reported by Kuo *et al.* [13].

During the transition from beam-on to beam-off, it is important that the time delay between the blanking signal at the upper and lower plates be equal to the electron transit time from the upper to the lower plates. The electron velocity (at 20 kV) is 0.27 times the velocity of light, and this translates into a 0.5-nsec delay for the 40-mm separation of the upper and lower plates. This delay is accomplished with a 50-Ω transmission line shown schematically in Fig. 14. In this, the beam passes between the horseshoe-shaped transmission line and the ground plane. Because the line is not balanced and not symmetric, there are current paths through the ground plane that differ from those in the strip; however, the unwanted beam deflection resulting from these currents is negligible.

The blanking of the beam would be complete only if all of the current at the intermediate image were confined within a region a few micrometers in diameter. Unfortunately this is not the case, and there is a measurable flux as far as 0.5 mm from the center of the image. When the beam is blanked by being deflected 8 μm onto the knife edge, the transmitted current falls by a factor of $\sim 10^5$ and is spread into a 1-μm^2 area. A sufficient current remains to cause an unwanted exposure equal to 1% of the normal exposure if the substrate is held still for only 128 μsec. To alleviate this problem, an additional set of deflection plates is provided that serves to deflect the beam an additional 500 μm and thus reduce the unwanted electron current by another factor of 1000. Additional measures (e.g., raster scanning the spot or moving the stage) must be taken if the stage is to remain stationary for longer than 128 msec. This secondary blanker is designed to operate in a few microseconds.

6. Substrate Height Detector

The height of the substrate can be measured with an optical detector. Light from an infrared (820-nm) semiconductor laser is focused on the substrate, and the reflected light is focused on the interface between the two halves of a split silicon photodetector. The beam is incident on the substrate at an angle of 5° to the horizontal. Any variation in the height of the substrate will cause a shift in the position of the focus on the photodetector, and the difference between the two signals (normalized by their sum) is a measure of the height of the substrate. If the deviations from the nominal position are small, this method is insensitive to the angular tilt of the substrate. It is designed to work over a range of at least ± 32 μm.

7. Mechanical Design

EBES4's electron optical design dictates the characteristics of the column's mechanical structure. It must have high mechanical stability since, unlike most scanning electron microscopes (SEM), which demagnify their source 40–100 times, the EBES4 column operates at unity magnification. Therefore, the column structure had to be stiff enough to prevent motions of the source, relative to the final lens, greater than $\frac{1}{64}$ μm. This was accomplished by rigidly mounting all the electron optical components to a large, very stiff aluminum casting (see Fig. 11). The casting provides large pumping ports on two sides to ensure a good vacuum throughout the column and two valves to isolate the column from the work chamber and the electron gun.

The entire column is enclosed in a three-layer mumetal shield to minimize the effects of slow changes in the ambient external magnetic field. As a result, the beam is deflected only ~0.1 μm by a slowly varying field of amplitude 1 μT. The effects of 60-Hz fields of similar magnitude are negligible. In order to increase the tolerance to external fields, an active system is provided to sense the field and correct the resultant error. The ambient magnetic field is sensed in two mutually perpendicular directions, and the signals are used to drive currents into two small coils inside the magnetic shielding. The linear transformation between the sensed fields and the coil currents is determined experimentally and produced with analog multipliers.

8. Column Control System

The column control system is run by an LSI-11/23™ computer. It controls the current sources for the lenses, stigmators, and centering coils. It is connected to the control computer by a serial 9600 baud (RS-232C) line, and the column electron optics can continue to operate indefinitely without the control computer.

The column control system provides 18 highly regulated (1 part in 10^5) voltages and currents to the gun, lenses, and centering coils. As a result, they are optically isolated from the column control LSI-11, and each supply is enclosed in a separate metal box. The LSI-11 controller can adjust any of the voltages or currents independently, or it can adjust all of them as a function of any other. The latter is referred to as coupling, and it is a great convenience when adjusting the column's optical parameters. For instance, if a centering coil is to scan the beam in a given direction (usually parallel to one of the machine's axes), the "cyclotron" rotation of the lenses may require both the X and Y coils to be driven independently. The column LSI-11 and the coupling hardware allow these ratios to be set by the control computer, but all of the necessary transformations between current changes in every coil to all others are maintained and calculated by the LSI-11 and the coupling hardware.

The high-voltage power supplies for the electron gun are not under the control of either the control or the LSI-11 computer. This is an exception to the general philosophy of giving the control computer complete control of the system, and it was made in order to reduce the possibility of an unwanted interruption in the gun's operation. As an additional precaution, these power supplies, together with that for the gun ion pump, are connected to uninterruptible power supplies, which can continue operation for 15 min during a power failure.

D. Mechanical Systems

The mechanical system comprises the entire physical structure of the machine. It includes the loader and the robot as well as the mechanical support for the electron column components.

One interesting aspect of the mechanical design is that the stage is supported and driven with high-pressure oil. This approach was chosen (1) because of its compactness, (2) because the large force-to-mass ratio allowed good dynamic performance, and (3) because it could be designed without moving electromechanical or magnetic elements. Figures 11, 15, and 16 are schematic representations of the electron column structure, the overall machine, and the drive system and stage structure, respectively.

The mechanical parts of the EBES4 system can be separated into several major subsystems: supporting structure, magnetic shielding, drive system, loading chamber, and robot. In the design, several overriding considerations affected the design of all subsystems:

(1) *Magnetic effects.* The electron beam must not "see" varying magnetic fields, so magnetic shields had to be used to minimize the effects of changing external fields, and no magnetic materials were allowed to move relative to the column.

(2) *Thermal effects.* Because the accuracy of the system depends on the mask substrate's temperature being both known and constant, the mechanical structure's temperature must be controlled.

(3) *Vibration effects.* Because the electron column operates at unity magnification, the cathode must be mechanically stable with respect to the measuring system of the moving stage. The sources of mechanical vibra-

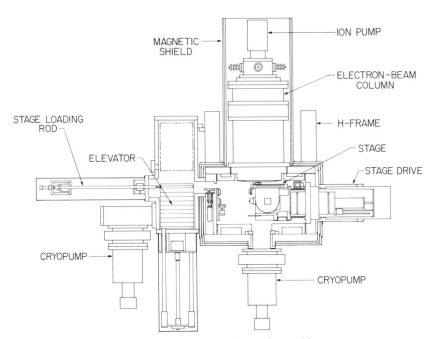

Fig. 15. Cross section of the entire machine.

tion on the machine must be minimized, and the entire structure must be stiff, thereby maximizing its resonant frequency.

(4) *Electronic effects.* Because the electron beam can be deflected by stray electric fields and because the precision of the electronics depends on a stable ground potential, the mechanical structure was not to be used for or connected to the electrical ground of any other system.

With these in mind, the mechanical system design grew around a massive bridge structure (see Fig. 15) that was supported on vibration isolation mounts. This structure supported the electron column and gun on top and the work chamber below. Both the drive system including the stage and the loading chamber are self-contained units that bolt to and form the vacuum seal for the work chamber. A three-layer magnetic shield surrounds the entire work chamber and electron column.

1. Supporting Structure

The supporting structure is dominated by two 4-in.-thick, 12-in.-deep aluminum beams. To these, the work chamber casting is attached, and the electron column is bolted to the top of the work chamber. Aluminum is used for all of these components. The work chamber is pumped by a single vibration-isolated cryopump mounted to the chamber bottom surface, and the column is pumped by one vibration-isolated cryopump and one Vac-Ion™ pump, both of which are supported by the bridge structure and not by the column.

2. Magnetic Shielding

The magnetic shielding surrounds the work chamber and column with openings for the loading chamber and vacuum pumps. The shield has three layers of mumetal separated by aluminum and plastic foam spacers.

Although the shield is made up of many sections, some with access doors, each of the shielding layers is continuous. The joining sections are designed to provide good magnetic and electrical connections between the respective shielding layers. The opening for the drive system is surrounded by a cylinder whose length-to-diameter ratio provides a good shielding factor even though this opening is 25 cm in diameter.

3. Drive System

The purpose of the drive system [14] is to support the mask substrate rigidly with respect to the interferometer mirrors and to move it in a single plane below the electron column. The motion must be smooth with controllable accelerations, and the drive system must not disturb the electron

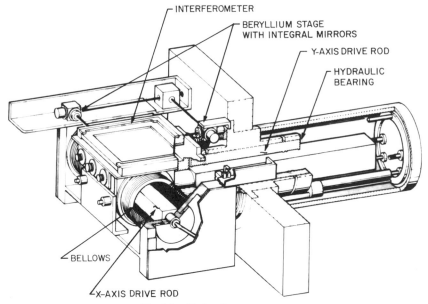

INTERFEROMETER

BERYLLIUM STAGE
WITH INTEGRAL MIRRORS

Y-AXIS DRIVE ROD

HYDRAULIC
BEARING

BELLOWS

X-AXIS DRIVE ROD

Fig. 16. Hydraulic stage system

beam either by creating mechanical vibrations that after being transmitted through the machine's structure vibrate the cathode or by having magnetic materials that move with respect to the beam.

Figure 16 shows the drive system. It is self-contained and can be withdrawn from the work chamber on a cart and replaced with a spare drive or operated outside the vacuum for testing purposes. When the drive has been installed in the work chamber, a three-layer magnetic shield covers the opening to complete the work chamber shielding. The drive system's major components are the following:

(1) *The stage.* The stage supports and rigidly clamps the cassette, which in turn supports the mask substrate. The outside edges of the stage have been nickel plated and polished flat to act as reference mirrors for the laser interferometers that measure the stage position. By making the mirrors an integral part of the stage and cassette-clamping mechanism, the interferometer will accurately measure the mask substrate position.

(2) *The "D".* The stage is supported on the X-axis drive commonly referred to as the "D" because of its shape. The "D" is hydraulically driven and supported on hydrostatic bearings.

(3) *The X-axis bearing rod.* A beryllium rod supports and guides the "D" along the X axis, and it is surrounded by two welded-metal bellows that form part of the hydraulic oil return and protect the work chamber

vacuum from contamination by the hydraulic oil. The interior of the bellows is evacuated by a roughing pump to minimize the differential pressure across the seals and to prevent the bellows from buckling due to excessive internal pressure.

(4) *The yoke.* The X-axis rod is supported on the magnesium yoke casting, and this in turn is supported on the Y-axis drive rod.

(5) *The Y-axis bearing rod.* As with the X axis, the Y-axis rod is beryllium and is supported and driven hydraulically, but in the case of the Y axis the rod moves and the surrounding bearing is stationary.

(6) *The Y-axis bearing and work chamber end plate.* The stationary bearing is attached to a thick plate that both supports the drive system and seals one side of the work chamber.

(7) *The hydraulic manifold.* The hydraulic manifold carries the servo valves for controlling the velocity of the X and Y axes, as well as filters, accumulators, and bypass valves to protect the drive system from excessive forces and to improve the performance of the drive system. This location of the valves, while not ideal for the system's dynamic performance, keeps their magnetic materials stationary with respect to the electron column.

(8) *The interferometer.* The Hewlett-Packard interferometers are attached to the work chamber plate to measure the Y-axis stage position and to the interferometer support rod to measure the position of the X axis. The support rod is fastened to the work chamber casting after the drive subsystem is installed to ensure that both interferometers are rigidly referenced to the electron column.

The oil used in EBES4 is not standard hydraulic fluid but 814Z™, a fluorocarbon oil developed for aerospace applications. Its salient characteristics are that it has an extremely low vapor pressure, and when it is struck by an electron beam, it does not carbonize, as normal hydrocarbon oils do, but cracks into gaseous products that can be pumped away. Therefore, in the event of an oil leak, this oil minimizes the risk of contaminating the electron column with a dielectric film that might be charged by stray electrons and deflect the beam. The oil's disadvantage is that it is more dense and has a lower bulk modulus than normal hydraulic oils, and this causes pressure waves to propagate more slowly through the 814Z™. The resulting 4-msec delay between the opening of the servo valve and the arrival of the pressure wave at the stage-drive cylinder complicated the design of the control algorithm.

4. Loading Chamber System

The loading chamber system provides a means by which cassettes containing mask substrates are introduced into the vacuum, onto the stage to be exposed, and then returned to the outside atmosphere. This must be done

(1) quickly, to minimize overheads,
(2) safely, so as not to endanger the rest of the system in the case of a malfunction,
(3) cleanly, so as not to contaminate the mask's surface with particles, and
(4) quietly (electrically and mechanically), so as not to introduce errors during the writing process.

The loading chamber and its vacuum system are under the direct control of the vacuum system LSI-11 microprocessor. The loading chamber is rough-pumped by a liquid-nitrogen-trapped mechanical pump, and final vacuum is achieved by a vibration-isolated cryopump.

Contamination of the mask plates is minimized by the following:

(1) The work chamber door is opened downward to eliminate the possibility of particles falling off the door or its seals onto the mask substrate, and sliding seals on the work chamber and loading chamber doors were avoided.
(2) Whenever the mask-holding cassette is moved, it is never allowed to slide. It is picked up and transported without contacting anything until it reaches its destination.
(3) The elevator slots are separated by septums to prevent debris from falling from one cassette onto the one below.
(4) The airflow rate into the loading chamber is limited to prevent a dust storm due to excessive air velocities.

5. Robot

The loading chamber is served by a Puma™ robot arm equipped with a vision system. It is capable of handling mask blanks and cassettes, and can load and unload the cassettes and transfer them between a storage area and the loading chamber. The vision system is used to check the operation at many stages and to determine the precise position of a mask blank that has been picked up with the vacuum chuck.

E. Control System

The control system provides the interface between the software that runs EBES and the electron column and mechanical systems. It gives the control computer a view into every aspect of the machine and allows it not only to monitor the status of every subsystem, but also to control every (controllable) variable. Although the control system components cannot be directly accessed (via knobs and switches) to influence the system's operation directly, various interfaces allow the operator to effect changes, and all such requests are passed through the control computer. Much of the control system consists of specially designed interfaces that process data and provide the "real-time" response that the control computer is too slow to provide.

In some cases the connections are simple and direct, such as loading or reading a register. In other cases they are complicated, and millions of operations are performed as a result of a single command, such as writing one stripe on every chip on a mask. Furthermore, the control system is complex because it must perform some functions for which the control computer is too slow and because it must be independent of the control computer for reliability reasons (as in the case of the vacuum control system as will be discussed in a later section).

The control system can be thought of as a collection of independent subsystems:

(1) *Stage control subsystem.* This subsystem accepts commands to position the stage and then positions the 256-μm beam deflection system to correct for stage errors. It also provides information to the fast-focus coil as a function of stage position.

(2) *Pattern memory control subsystem.* This subsystem coordinates the entire writing process using lists of instructions stored in its "pattern memory" to position the stage, to set the 32-μm deflector, and to write figures with the 4-μm deflector.

(3) *Loading chamber control subsystem.* This subsystem, operated by the vacuum control LSI-11 microprocessor, uses a robot to load mask substrates into cassettes, place them into the 10-cassette loading chamber, and then transfer them from the loading chamber to the stage and back.

(4) *Sensor subsystem.* This subsystem monitors all of the vacuum, position, force, pressure, temperature, and displacement sensors. It also digitizes the signals for presentation to the control computer or to other subsystems.

(5) *Column control subsystem.* This subsystem, discussed in Section IV.C, controls the high-voltage supplies for the electron gun and the precision current sources for the electron lenses, the stigmator coils, and the centering coils.

(6) *Vacuum control subsystem.* This subsystem monitors (through the sensor subsystem) and controls all of the vacuum system's pumps and valves.

1. Stage Control Subsystem

The stage control subsystem receives its instructions in the form of WYWHs in metric coordinates and then calculates the corresponding position in machine coordinates, taking into account the laser interferometer wavelength and known distortions of the mechanical system. The error in the stage position provides outputs to the stage drive servo system, which positions the stage, and positions the 256-μm beam deflection system to correct for these errors.

The WYWHs assume an orthogonal coordinate system with straight axes and a plane writing surface. Unfortunately, imperfections in the mechanical system would reduce the absolute accuracy of the final product were it not for the control system's capacity to compensate for these imperfections. Two separate control algorithms correct for errors in the horizontal plane (the X and Y directions) and in the vertical (focus) direction. In the horizontal plane the corrections are increments added to or subtracted from the nominal WYWH location to account for the fact that the measuring system (the interferometer) operates in fractional wavelengths of light and not in micrometers and that the coordinate system (defined by the stage mirrors) is neither straight nor orthogonal. In the focus direction (Z), the vertical position of the writing surface is measured at several locations by an optical technique (using an infrared laser system described above), and a smooth surface is fitted through these points. During writing, the vertical position of the writing surface is continually calculated, and the result is used to correct the focus and to make corrections to the gains and rotations of the magnetic electron-beam deflection system made necessary by the focus change. The algorithms for the $X-Y$ and Z corrections are similar, so only the first is illustrated below.

We transfer the "nominal" WYWHs X and Y into machine coordinates using the following formula:

$$X_{mach} = X_{nom} + \alpha_1 + \alpha_2 X_{nom} + \alpha_3 Y_{nom} + \alpha_7 X_{nom} Y_{nom}$$

$$Y_{mach} = Y_{nom} + \alpha_4 + \alpha_6 Y_{nom} + \alpha_5 X_{nom} + \alpha_8 X_{nom} Y_{nom}$$

In most cases, the distortions of the machine or substrate cannot be corrected with one set of α coefficients representing the entire writing area. In these cases the stage control hardware allows a complex error "surface" to be approximated by up to 1000 small "alpha cells," each of which may have a different set of α coefficients.

One interesting potential application of EBES4's capacity to use many "alpha cells" is reticle predistortion to compensate for the lens distortion in direct step on wafer (DSW) step-and-repeat cameras. If the lens distortion

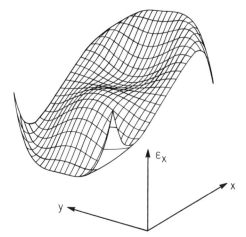

Fig. 17. Example of the distortion corrections that can be applied with the stage control system. The X error (plotted vertically) over the $X - Y$ plane will correct the third- and fifth-order distortion of the lens of an optical step-and-repeat camera. A similar correction must be applied to the Y direction.

in a particular camera is too large, the same camera must be used for every lithographic step to ensure that the patterns overlay properly. Alternatively, EBES4 could be used to write reticles with a compensating error so that the resulting image on the semiconductor surface is distortionless. As a result, the distortion "signature" of a DSW could be removed and subsequent lithographic steps could be performed on any DSW. As an example, Fig. 17 shows the X error (plotted vertically) over the $X - Y$ plane for a lens with third- and fifth-order pin-cushion distortions of opposite signs. The use of this technique is not thought likely at present because of the very high quality of DSW lenses and the large expense of providing a different set of masks for each DSW for every pattern.

2. Pattern Memory Control Subsystem

The pattern memory control subsystem (Fig. 18) coordinates the entire writing process. The control computer loads the pattern memory with the description of a portion of the mask (usually the description of all figures in one stripe of every chip and the location of every chip), and when the subsystem is started by the control computer, it coordinates the entire writing process until that portion of the mask is written. In doing this it expands the data, which are stored in a compact form, and routes the data in the correct order to the appropriate subsystems: (1) stage control and

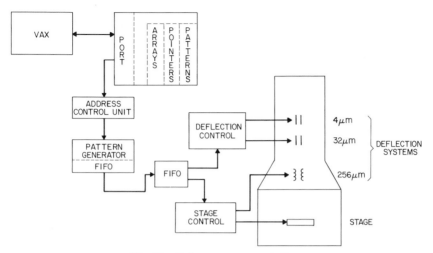

Fig. 18.   Pattern memory controller.

256-μm deflection system, (2) 4-μm deflection control, (3) 32-μm deflection control, and (4) SEM deflection control. The result is a system made up of a large random-access "pattern memory" (PM), a series of special-purpose processors that address the PM and expand the resulting data during the writing process, an interconnection bus, and a series of deflection systems that control the beam position during writing.

a. Pattern Memory.   This 4-Mbyte random-access memory is dual ported. It can be written and read by the control computer, and it can be read by the address control unit (ACU), the first of the special-purpose controllers. The dual-port logic allows both the control computer and ACU to access the memory, and it resolves simultaneous requests for the memory. The memory normally stores all of the information required to write one stripe of all the chips on the mask. These data are stored in three variable-length tables (array, pointer, and pattern), and as the memory is read by the ACU during the writing process, the control computer can be loading data into the memory for the next stripe of the pattern. The contents of the three tables are as follows:

(1)   *Array table.* This describes the X and Y location of each chip's centerline and the starting location in PM where the "pointer table" description of that chip begins. It may also contain the nominal stage drive speed (called feed forward) at which the chip is to be written.

(2)   *Pointer table.* This describes the location of each 32-μm cell rela-

tive to the chip's centerline (called cell offset) and the location in PM where the "pattern table" for that cell begins. The pointer table contains both the X and Y offset of the cell and the desired X and Y offset of the stage position from the chip's centerline.

(3) *Pattern table.* This describes the type and location of each of the figures in the cell. Locations are given relative to the cell's origin, and there are seven basic figure types: a rectangle, four orientations of the 45° right triangle, and two types of the 45° parallelogram.

b. Address Control Unit. This special-purpose processor controls the order in which instructions are read from the PM. The control computer gives the ACU a starting address in the PM, and the ACU begins fetching sequential words starting at that address. As words are fetched, the ACU interprets each word. If the word does not contain an instruction for the ACU, it is passed onto the next processor, the pattern expansion unit. If the word is an instruction for the ACU, it uses the instruction to calculate the address of the next word and begins fetching sequential words starting at the new address.

During the writing process, words are fetched from PM in sequence until it is time to transfer from one table to another. This is accomplished with "jump subroutine" (JSR) and "return" type of instructions. These instructions allow sections of the pointer and pattern tables to be called repeatedly as though they were subroutines. This minimizes the pattern memory size by requiring the storage of only a single occurrence of each chip description or repeated subpattern within the chip. In addition, the pointer table may be read forward from beginning to end or backward. This allows chips to be written in either their normal or mirrored orientation or to be written from either right to left or left to right using the same description.

c. Pattern Expansion Unit (PEU). This special-purpose processor, like the ACU, reads and interprets instructions intended for itself and passes the other instructions on to the next processor, the FIFO (First-in first-out buffer) and ECL (emitter coupled logic) bus. The purpose of the PEU is to expand the instructions that describe the large (up to 32-μm) "macrofigures" into a series of small (up to 2-μm) "microfigures" that completely tile the surface of the original macrofigure. It expands each macrofigure description at the rate of one microfigure per 100 nsec. Each of the microfigure descriptions contains the figure type, size, and displacement to the succeeding figure. These microfigure descriptions and the deflection system and stage control commands are transmitted by the ECL bus to the machine-mounted electronics that interpret and carry out the commands.

3. Loading Chamber Control Subsystems

The steps in the EBES writing process include selecting an unexposed mask substrate, loading it into a cassette (this is a mask substrate holder that adapts plates of various sizes to the stage-clamping mechanism), putting the cassette into the loading chamber, evacuating the loading chamber, opening the door between the loading chamber and work chamber (which always remains evacuated), transferring the cassette to the stage, exposing the substrate, and then reversing the sequence to retrieve the exposed substrate. The loading chamber control subsystem controls the robot that loads glass into the cassette and later unloads it, and it controls the operations within the loading chamber. The sequence of operations is as follows:

(1)   The robot is directed to transfer an unexposed plate from a rack containing an assortment of unexposed plates to a cassette of the correct type. A robot is used to minimize both the thermal and the particulate contamination of these mask substrates.

(2)   The robot then transfers the loaded cassette to the loading chamber's transfer station.

(3)   Up to 10 cassettes are transferred, one at a time, from the transfer station to a 10-position rack in the loading chamber by a transfer arm on the loading chamber.

(4)   The loading chamber door is closed, and the chamber is evacuated.

(5)   Cassettes are sequentially transferred between the loading chamber and the stage in the work chamber where they are exposed.

(6)   After exposure, the process is reversed and each cassette containing an exposed plate is removed from the loading chamber.

(7)   The robot removes the exposed plate from the cassette and puts it in the "to-be-developed" box.

The robot was incorporated into EBES4 to minimize particulate contamination (by avoiding human presence) and to minimize thermal disturbances around the machine when handling masks.

4. Sensor Subsystem

The EBES4 system is monitored by dozens of sensors to ensure its safety and proper operation. These include the following:

(1)   Temperature sensors in the column, the hydraulic drive system, and the loading chamber

(2)   Pressure sensors in the vacuum control system and the hydraulic drive system

(3) LED limit switches on all moving parts

(4) Capacitance proximity detectors for measuring cassette seating on the stage

(5) Vacuum sensors

(6) Voltage and current sensors for the beam current and the power supply voltages

Most of the signals from these sensors are conditioned by electronics mounted on the machine to boost the signal levels. The analog-to-digital conversion for most of the signals is performed by the vacuum control LSI-11 because many of the signals are needed by this microprocessor to control the vacuum system safely.

5. Vacuum Control Subsystem

The vacuum control system measures the pressures throughout the column, work chamber, drive system, and loading chamber, and it controls all of the vacuum valves. Most of these valves are controlled pneumatically via electrically operated solenoid valves. Great care was taken in the design of the pneumatic control system and the LSI-11 software to protect the vacuum from most accidents and hardware failure modes.

F. Software

The EBES control program provides the overall control of the machine as well as the interface between the machine and the operator and between EBES4 and the mask shop information and management system (MIMS). In addition, there are maintenance and diagnostic programs used by the developers to debug and test the mechanical, electronic, and electron optical systems, and many of these will be used in the future by mask shop personnel to monitor the machine's health. The software falls into several usage categories:

(1) *The EBES4 writing program.* Unlike the software of previous EBES systems, the EBES4 program combines the pattern writing, system monitoring, loading chamber control, column alignment, deflection alignment, MARKET (a mask-measuring program), and MIMS communication routines into one unified program.

(2) *The pattern data preparation (preprocessing) programs.* Although the pattern preprocessing programs can run on the EBES computer system, we plan to transfer them to the mask shop, so that the pattern data will be prepared on mask shop computers before sending the job to EBES4.

(3) *The maintenance and debugging programs (MAINT).* The

MAINT language was written to assist hardware developers to communicate with their hardware during the debugging and testing phases. In addition, several special-purpose programs were written to test and exercise the pattern memory and its associated processors (ACU, PEU, ECL bus).

The software was written and is maintained on the EBES system's VAX control computer. Most of the software is written in FORTRAN 77 for both this computer and the LSI-11's. The latter use a monitor program that allows programs written, compiled, and loaded on the control computer to be down-line-loaded over an RS-232 serial line into the LSI-11 and then run. Virtually all of the LSI-11 hardware and software debugging has been done via a VAX/VMS™ terminal through the VAX computer's serial link to the LSI-11. In this case, the control computer appears to both the LSI-11 and the programmer as a terminal, but it provides the additional advantage of having available mass storage, hard copy, and many software tools. This approach eliminated the need for several LSI-11 development systems and eliminated the inevitable compatibility problems that arise when the software is moved from the development system to the EBES4 computer.

EBES4 can be described as a "software machine" because the system philosophy, set down early in the machine's development, was that all controllable and readable variables in the hardware would be controlled and read by the control computer and that the operator would not have direct control of any machine functions without control computer support. Therefore, there are no knobs or switches that directly control power supplies (except for the electron gun high-voltage supplies). The machine, however, does have some special features to make operator intervention easy:

(1) The operator console (a standard VT100 terminal) has been augmented with a touch-sensitive screen, and software utilities have been written to format the screen with outlined, labeled "buttons." However, when it is convenient to run the system from another terminal (which may be a great distance away), rather than a touch screen being used the commands can be typed. These commands can also be read from script files stored on the disk, and the files can be nested.

(2) The command language provides a wide range of facilities, including timers and conditional branching. The timers can be used to schedule tasks that will be run unattended.

(3) The operator can "attach" additional display consoles (to the EBES4 program to provide continuously updated displays of alpha numeric status information) and a color graphics display. As with the buttons, utilities have been written to allow the user to create new status displays.

Frequently used displays include vacuum status, system temperatures, column status, and drive system status.

(4) The "knob box" allows the operator to change variables that are under the control of the computer with a familiar twist of a knob. This peripheral device contains eight knobs and an alphanumeric display that is used to label the knobs. Each knob is connected to a shaft encoder, and as the knob is turned, a microprocessor interprets the encoder's output. Periodically, a message is sent to the control computer over an RS-232 serial line indicating which knob was turned, how far, and in which direction. The control computer interprets this as a command and performs the required function. These routines are so fast that an operator actually feels as though the function is being controlled directly. Utilities have been written to program and label the knobs. This is particularly valuable because some of the functions are complex, such as controlling two voltages at the same time or controlling the ratio between two voltages.

The EBES control program is built around a central command-queue–parser utility. Figure 19 shows how this operates. All commands from the operator's console, the touch screen monitor routine, command lists, or other EBES4 processes are parsed and interpreted, and the action routines are called to perform the desired function. Notice that there are several queues, each with a different priority, onto which commands can be

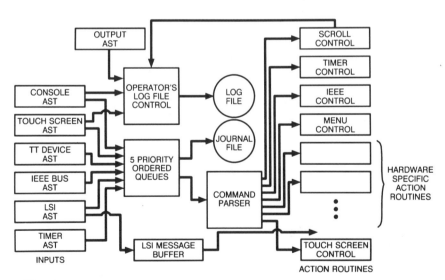

Fig. 19. Structure of the EBES4 control program. AST: a VAX/VMS asynchronous system trap; IEEE: the IEEE-488 bus; LSI: LSI–11/23 computers; TT: terminals other than the main console terminal.

placed. This allows high-priority commands to be executed immediately, eliminating the need to wait in line behind low-priority commands. In addition to starting the action routines, the parser sends the commands to a log file in the order that they were parsed. The log file may be reviewed by the developers after a hardware or software malfunction to unravel the sequence of events leading to the malfunction.

The use of a command queue has alowed the rapid development of application programs within the structure of the EBES4 program by minimizing the number of system-level details that must be considered by the applications programmer. It has meant that both software and nonsoftware developers have been able to build the system software and application tasks without having to understand the software system's entire structure. On the negative side, it has imposed restrictions on the operator – action-routine communication and it has increased the complexity of the action routines.

Now that we have described the essential elements of the EBES4 program, several other functions deserve more attention: the writing routines, the column alignment routines, and the deflection alignment routines.

The writing routines (1) accept requests to write a mask from MIMS, (2) organize the data, (3) control the system during the writing process, and (4) return the writing results to MIMS at the completion of the job. In more detail, the sequence of events is as follows:

(1)   A job list is received from MIMS giving the masks to be written.

(2)   The robot or an operator is requested by the EBES4 program to load an appropriate unexposed mask blank (correct size thickness, glass type, chrome type, resist type) into a cassette and the cassette onto the loading chamber transfer station.

(3)   Assuming that this is the only job, the loading chamber control routine loads the cassette onto the stage.

(4)   The mask array information is read, and the necessary column and deflection system parameters are extracted. Then the column and deflection system alignment routines are called (more about these later).

(5)   The mask substrate is scanned with the laser focus detector, and appropriate information is sent to the stage control's focus-control algorithm. The algorithm will use the information and the current stage position to calculate the fast-focus coil current needed to keep the beam in focus.

(6)   The fiducial marks are found to establish the absolute relationship between the positions of the electron beam and the stage.

(7)   The alphanumeric mask identification information is written using routines in PM.

(8)   The step-and-repeat array information is read, and the pattern information for all of the required chips is read, sorted suitably, and loaded into PM.

(9)   The control of the electron column and stage drive is turned over to the PM control subsystem. While writing is taking place, the information for the next stripe is assembled by the control computer in vacant PM space. Between stripes, control returns to the control computer, where it checks for the successful completion of the previous stripe before starting the writing of the next stripe. This process is repeated until the mask is complete.

(10)   The loading chamber control routine is requested to remove the cassette, and MIMS is given the statistics of the writing process.

A very important but invisible part of the writing routine is the recovery from errors. The above description assumed that every activity was completed as requested. However, errors will sometimes occur (some recoverable and some not); the program must not "blow up," but must attempt to recover or terminate the job and communicate something intelligible to the operator so that corrective actions can be initiated.

To prepare for writing a mask, another set of routines is used to adjust the column's electron optical parameters. The tasks include the following:

(1)   Centering the first lens.

(2)   Centering the beam on the beam-limiting aperture. This operation maximizes the beam current through the aperture.

(3)   Adjusting the current of the first lens to set the focal length of the lens and thereby controlling the total beam current through the limiting aperture.

(4)   Adjusting the stigmator coil current to make the beam round where it focuses at the beam-blanking knife edge.

(5)   Adjusting the second and third centering coils to position the beam relative to the beam-blanking edge and to establish its direction parallel to the column's centerline. Because the blanker deflects the beam only a few micrometers, these adjustments are crucial to achieving an acceptable on/off ratio.

(6)   Adjusting the current of the second lens to make the beam focus at the beam-blanking knife edge.

(7)   Adjusting the current of the third lens to focus the beam on the resist-coated surface of the mask.

(8)   Adjusting the fourth centering coil to center the beam in the final lens and thus minimizing off-axis aberrations.

(9)   Adjusting the second stigmator to make the beam round on the mask's surface.

Each of these steps is made by first measuring some parameter of the beam (size, position, current) at either the beam-blanking knife edge or at the writing plane and then making a change in the required parameter and repeating the measurement. Under software control the column alignment requires only minutes instead of the hours required to perform a complete alignment by hand. In addition, the results are as good as or better than can be achieved by a skilled engineer.

Finally, the deflection-alignment routines of the EBES4 program adjust the direction (called rotation) and magnitude (called gain) of the beam deflection to coincide with the machine's axes and to agree with the laser interferometer. The deflection systems that must be aligned are

(1)   256-μm magnetic deflection — gain and rotation,
(2)   32-μm electrostatic deflection — gain and rotation, and
(3)   4-μm electrostatic — gain.

In the case of the 256-μm deflection system, its deflection must be precisely aligned to compensate for stage-position errors. As mentioned in Section IV.E, the WYWH location is the position on the stage where the electron beam should hit (assuming no 32- or 4-μm deflection), and this is achieved by moving the stage to the approximately correct location and then by deflecting the beam to correct for the stage location error. The error budget allows less than $\frac{1}{64}$-μm error due to improper deflection alignment.

The 256-μm deflection-gain and deflection-rotation measurements are made by using the electron-beam deflection system to determine the location of a single fiducial mark that is placed at many points in the 256-μm deflection field by moving the stage. If the relative locations of the marks, as measured, agree with the relative stage positions (as measured by the interferometer) at which the marks were positioned, the alignment is correct. If not, the gain and rotation coefficients are altered and the process is repeated.

The 32- and 4-μm deflection systems are adjusted in a similar manner but, as their names indicate, over a reduced field size.

V. CONCLUSIONS

We have traced the evolution and development of electron-beam lithography equipment for mask making at AT&T Bell Laboratories and described EBES4, our latest system. EBES4 represents a step forward in the accuracy and resolution of our pattern generators with its $\frac{1}{8}$-μm writing spot and $\frac{1}{16}$-μm minimum address size. At the same time, it has a greater throughput than any of the existing pattern generators based on EBES II. Its greater degree of automation in system setup and mask handling will allow remote and unattended operation and reduce the level of operator skill required for routine use.

REFERENCES

1. D. R. Herriott and G. R. Brewer, *in* "Electron-Beam Technology in Microelectronic Fabrication" (G. R. Brewer, ed.), pp. 141–216. Academic Press, New York, 1980.
2. J. P. Ballantyne, *in* "Electron-Beam Technology in Microelectronic Fabrication" (G. R. Brewer, ed.), pp. 259–307. Academic Press, New York, 1980.
3. M. G. R. Thomson, R. J. Collier, and D. R. Herriott, *J. Vac. Sci. Technol.* **15,** 891–895 (1978).
4. G. Cogswell, S. Miyauchi, T. Tanaka, and N. Goto, *Proc. Int. Conf. Electron Ion Beam Sci. Technol., 8th* pp. 117–134. Electrochem. Soc., Princeton, New Jersey, 1978.
5. R. D. Moore, G. A. Caccoma, H. C. Pfeiffer, E. V. Weber, and O. C. Woodard, *J. Vac. Sci. Technol.* **19,** 950–952 (1981).
6. H. J. King, P. E. Merritt, O. W. Otto, F. S. Ozdemir, J. Pasiecznik, A. M. Carroll, D. L. Cavan, W. Eckes, L. H. Lin, L. Veneklasen, and J. C. Wiesner, *J. Vac. Sci. Technol., B* **3,** 106–111 (1985).
7. J. C. Eidson, W. C. Haase, and R. K. Scudder, *Hewlett-Packard J.* **32**(5), 3–13 (1981).
8. D. S. Alles, C. J. Biddick, J. H. Bruning, J. T. Clemens, R. J. Collier, E. A. Gere, L. R. Harriott, F. Leone, R. Liu, T. J. Mulrooney, R. J. Nielsen, N. Paras, R. M. Richman, C. M. Rose, D. P. Rosenfeld, D. E. A. Smith, and M. G. R. Thomson, *J. Vac. Sci. Technol. B* **5,** 47–52 (1987).
9. M. G. R. Thomson, R. Liu, R. J. Collier, H. T. Carroll, E. T. Doherty, and R. G. Murray, *J. Vac. Sci. Technol. B* **5,** 53–56 (1987).
10. F. M. Charbonnier, E. E. Martin, and L. W. Swanson, Field emission cathode having tungsten Miller indices 100 plane coated with zirconium, hafnium, or magnesium on oxygen binder, U.S. Pat. 3, 374, 386 (1968).
11. J. E. Wolfe, G. E. Ledges, and H. H. Glascock, Electron emission system, U.S. Pat. 3,814, 975 (1974).
12. R. Liu and M. G. R. Thomson, Electron emission system, U.S. Pat. 4, 588, 928 (1986).
13. H. P. Kuo, J. Foster, W. Haase, J. Kelly, and B. M. Oliver, *Ext. Abstr. Int. Conf. Electron Ion Beam Sci. Technol., 10th* p. 481. Electrochem. Soc., Princeton, New Jersey, 1982.
14. R. J. Nielsen, J. H. Bruning, R. M. Richman, C. J. Biddick, J. Giacchi, G. J. W. Kossyk, D. R. Bush, S. J. Barna, and D. S. Alles, *J. Vac. Sci. Technol. B* **5,** 57–60 (1987).

Chapter **4**

Electron Resist Process Modeling

N. EIB

General Technology Division
IBM Corporation
Hopewell Junction, New York 12533

D. KYSER

Philips Research Laboratories
Cignetics Corporation
Sunnyvale, California 94088

R. PYLE

Data Systems Division
IBM Corporation
Poughkeepsie, New York 12602

I. INTRODUCTION

The technology of electron-beam lithography (EBL) depends on the interaction of a focused electron beam with a polymeric resist film or

103

multilayer film structure on a nonpolymeric substrate. In general, electron irradiation of a polymeric film produces microstructural changes such as polymer chain scission in positive resists or polymer chain cross-linking in negative resists. Such changes produce patterns in the resist film with differing solubility rates that correspond to the pattern of the original electron-beam exposure pattern. Selective solvents are then utilized to "develop" the film for subsequent use as a resist mask for chemical etching of the substrate, ion implantation, and other high-resolution microfabrication processes. EBL, x-ray, and ion-beam lithography are being utilized to generate submicrometer structures in advanced device research and development programs.

The spatial contour or resist profiles developed in the resist film are determined by at least two independent processes: (1) electron scattering and energy deposition within the film–substrate target and (2) the specific development process of the electron-irradiated volume with a solvent. Due to the large number of relational parameters encountered in these two processes, it is often necessary to use simulation as a tool to understand and quantify these parameters in the context of the overall lithography process. For a particular process application, it is important to optimize or trade off such parameters as electron-beam voltage and shape, electron exposure (dose), multilayer resist film thicknesses, solvent concentration and temperature, and development time to achieve a desired resist profile or process tolerance (or both).

In particular, it is important to understand *quantitatively* and compensate for the well-known electron scattering effects, known as "proximity" effects, unique to EBL. Proximity effects are due to electron scattering within the film–substrate target. Electron scattering that originates from within a pattern feature is defined as intraproximity, and scattering that originates from adjacent features is defined as interproximity. The total energy deposited within a pattern feature depends on both its own size (intraproximity) and the size and distance of neighboring pattern features (interproximity). As pattern feature sizes and spacings change, proximity effects become more or less influential. One can compensate for these proximity effects via modulation of the exposure dose or pattern size.

The art and science of EBL has already received much attention in the scientific literature (see Thornton [1,2] and Munro [3] for a general review of EBL). Some early work on process modeling is described by Hawryluk *et al.* [4], Greeneich and Van Duzer [5], Kyser and Murata [6], Neureuther *et al.* [7], Kyser and Pyle [8], and Chang *et al.* [9]. A review of exposure and development models is provided by Hawryluk [10]. Some three-dimensional modeling of pattern features is described by Jones and Paraszczak [11], and a critical study of tool–resist interdependence is described by Eib and Jones [12].

In this chapter we describe computer simulation with experimental corroboration of electron-beam exposure, resist development, and the resulting resist profiles. The unique properties of a nonlinear resist [8,13] are utilized as an example of current technology, and various approaches to proximity correction are discussed and critiqued. A detailed method of characterizing the development process with such nonlinear resists is also described (see Eib [29]).

II. ELECTRON-BEAM EXPOSURE MODELING

The interaction of a focused electron beam with a resist film – substrate target has been approximated by a variety of mathematical models. The models are either analytic, closed-form solutions of statistical, Monte Carlo based solutions. In both cases, approximations are utilized to generate a quantitative value for the spatial distribution of energy density E_v (eV/cm^3) deposited within the resist film. The incident electron beam has a primary voltage E_0 (keV) and imparts an exposure dose Q (C/cm^2) for a two-dimensional areal source or exposure dose q (C/cm) for a one-dimensional line source. Due to electron scattering within the target and to nonlinear rates of energy loss per unit electron path length, there is a spatial distribution of E_v that depends on beam accelerating potential E_0, resist film thickness X_0 (μm), substrate atomic number Z, and substrate thickness. If the substrate is sufficiently thin, primary electrons can traverse the substrate without losing all their kinetic energy within the resist – substrate target. The spatial distribution of absorbed energy in the electron resist is referred to as the latent image.

In this chapter, we utilize a Monte Carlo method for calculating the trajectories of primary electrons within a target such as that utilized in EBL (see Hawryluk [10] for analytic, non-Monte Carlo methods). Several recent papers have described in detail the Monte Carlo calculation methodology, including those of Kyser [14] and Murata [15]. The "single-scattering" Monte Carlo model utilized in this chapter is described elsewhere [14,15].

A. Electron Trajectories and Energy Density Contours: Bulk Targets

Within the Monte Carlo methodology, the spatial distribution of energy density E_v deposited by an electron beam is actually generated as a histogram in half-space subdivided into cells. The spatial resolution of the histogram cell is preset to some value (e.g., 0.05 μm) and then a large number of electron trajectories (e.g., $10^4 - 10^6$) are calculated to provide sufficient statistical precision for the analysis. This is necessary because no

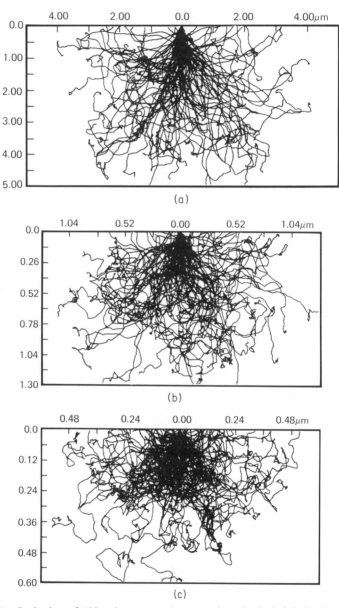

Fig. 1. Projection of 100 point source electron trajectories in infinitely thick (from an electron scattering point of view) targets of (a) silicon ($\rho = 2.33$), (b) copper ($\rho = 8.96$), and (c) gold ($\rho = 19.32$) at an incident energy of 25 keV. The x and y axes are in units of weight per unit area (g/cm^2).

two electron trajectories are identical. Pseudorandom numbers are generated to determine scattering angles, weighted probabilities of atomic interaction, path length between collisions, etc. The energy deposited in each cell volume via Rutherford scattering is accumulated from each electron that traverses that cell and converted to an average energy density E_v for each cell. The calculation can be made for a point source beam, a line source beam (which is the same as a point source scanned in one direction), a shaped beam (i.e., a projected feature exposed simultaneously), or a plane source beam (i.e., a uniform areal exposure of the whole resist film). Each type of calculation has specific applications for modeling, either for profile simulation or for resist development characterization, to be described in Section III.

An example of the Monte Carlo results obtained for a wide variation in target material is shown in Figs. 1 and 2. For visual clarity, only 100 electron trajectories are shown. Because of the strong dependence of elastic scattering and energy loss on atomic number Z, the targets of silicon ($Z = 14$), copper ($Z = 29$), and gold ($Z = 79$) exhibit quite different trajectories. Note that a scale in units of mass thickness $\rho x (\mathrm{g/cm^2})$ has been utilized to normalize the differences in mass density ρ $(\mathrm{g/cm^3})$. This is valid since the mean free path for elastic scattering scales linearly with ρ [14,15]. For statistical precision, each of the three substrates was simulated with 10^5 electrons. Histogram cells with the same absorbed energy density were then connected graphically to produce the equal energy contour plots shown in Figs. 2a–c. It is apparent that the shape and position of the contours change dramatically with Z. This has direct effects on electron resist films placed on top of such substrates and will be discussed in the next section.

B. Electron Trajectories and Energy Density Contours: Film–Substrate Targets

When an electron resist film is placed on top of a substrate target, the influence of the resist material and thickness must be included in any model of electron lithography processes. The discontinuity of the electron scattering and energy loss at resist–substrate interfaces is easily described within a Monte Carlo model *without adding any additional models.* Such is not the case with analytic models, and this is why the authors continue to favor the Monte Carlo method for general application in electron lithography (see Rishton [30] for a discussion of analytic limitations). As an example of the case in which a polymeric film is exposed on a substrate, Figs. 3 and 4 show 100 electron trajectories at 10, 25, and 50 keV incident on a 1.0-μm resist film on silicon and gold substrates, respectively. The

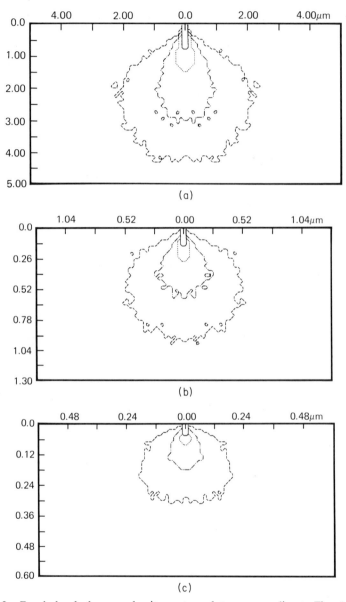

Fig. 2. Equal absorbed energy density contour plots corresponding to Figs. 1a–c. The relative absorbed energy density ratios (RATO) are 16.00, 8.00, 4.00, and 2.00. The highest value corresponds to the innermost contour.

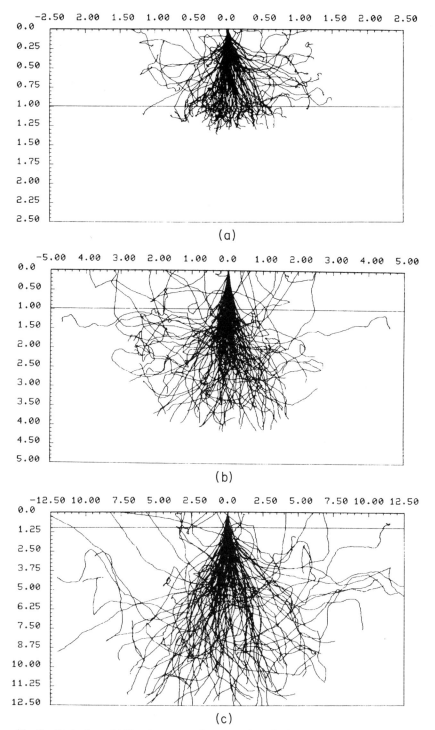

Fig. 3. Projection of 100 point source electron trajectories in a target of 1-μm-thick resist on an infinitely thick silicon substrate at (a) 10, (b) 25, and (c) 50 keV incident energy. The x and y axes are in units of micrometers.

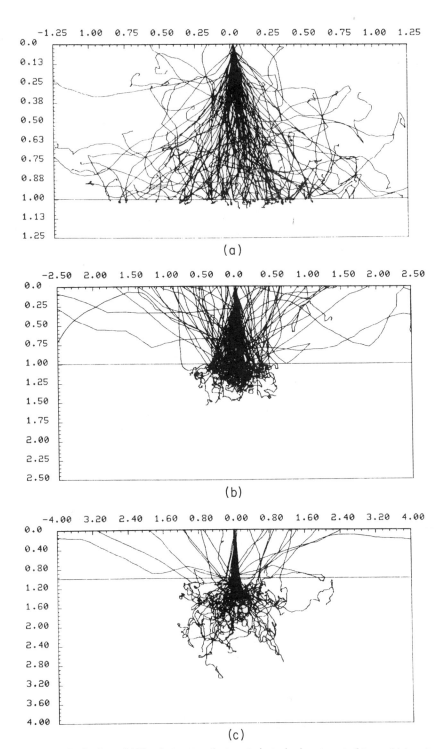

Fig. 4. Projection of 100 point source electron trajectories in a target of 1-μm-thick resist on an infinitely thick gold substrate at (a) 10, (b) 25, and (c) 50 keV incident energy. The x and y axes are in units of micrometers.

resist film is assumed to be equivalent to poly(methyl methacrylate) (PMMA) with a weight percent composition of 60% carbon, 32% oxygen, 8% hydrogen, and a density of 1.22 g/cm³. The Monte Carlo model automatically includes scattering by a compound chemical target with specified composition. Such trajectories are representative of those in most organic materials. Note that the scales in Figs. 3 and 4 are different. An important factor for the latent image and proximity effects is the maximum lateral range of scattered electrons. The lateral range increases with increasing beam voltage and decreases with increasing atomic number, while the fraction of electrons backscattering from the substrate increases with increasing atomic number. This has consequences for proximity effects and their correction or compensation. The effects on developed images will be discussed in Section IV.

Some insight into the effects of beam voltage and film thickness on absorbed energy density can be gained from calculations of a uniform areal exposure of the whole resist film. Such effects are demonstrated in Fig. 5 for six values of beam voltage (5 – 50 keV) and five values of resist film thickness (0.5 – 1.8 μm) on an infinitely thick silicon substrate (infinitely thick from an electron penetration point of view). The curves in Fig. 5 are derived from a linear least-squares fit to a one-dimensional histogram generated via Monte Carlo calculations with a 0.05-μm cell size. Note that each curve ends at the appropriate film thickness, namely, the resist–substrate interface. The maximum penetration is 0.5 μm at 5 keV and 1.8 μm at 10 keV. The energy density has been normalized to the areal exposure Q for ease of comparison. These results are similar to those

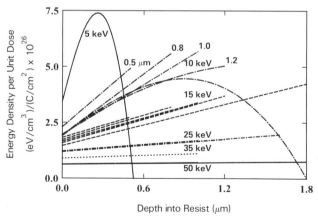

Fig. 5. Absorbed energy density per unit dose as a function of depth in resist films with thicknesses ranging from 0.5 to 1.8 μm on an infinitely thick silicon substrate at incident energies ranging from 5 to 50 keV for a plane (blanket exposure) source.

obtained previously by Kyser and Murata [6] and illustrate the strong dependence of resist "sensitivity" on beam voltage but relatively weak dependence on film thickness. The resist sensitivity is properly described only by a development rate (or rate ratio) and a relationship between development rate and absorbed energy density. This relationship is discussed in Section III.

C. Proximity Effects in Latent Images

The dependence of energy density deposited in a finite linewidth $W(\mu m)$ for constant resist film thickness and substrate is shown in Fig. 6. These curves correspond to the latent image cross section for an infinitely long line. The lines are deliberately separated to avoid interproximity effects. The solid, dotted, and dashed lines represent, respectively, the top, middle, and bottom 0.05-μm-thick layers in a 1.0-μm-thick resist film on infinitely thick silicon. This illustrates the gradient in the latent image with depth. The lateral shape of the curves illustrates the effects of electron scattering after the incident beam enters the target for line widths W of 1.0, 2.5, and 5.0 μm. The "edge width" of the beam (see Kyser and Pyle [8]) is 0.1 μm for all simulations in this chapter. In a real electron-beam tool, the edge width may vary with linewidth in a shaped beam system.

Several important characteristics of electron scattering are shown in Fig. 6. First, only intraproximity effects are illustrated. At 10 keV, the gradient in the latent image with depth into the resist is large, compared with 25 and 50 keV. This arises because of the relatively strong lateral scattering of electrons *within the resist film* at lower beam voltages (see Figs. 3 and 5). However, the intraproximity effect tends to saturate at narrower linewidths at 10 keV, compared with higher beam voltages. Second, the total electron scattering range at the higher beam voltages results in exposure of the resist at all depths far from the edges of the incident beam. These "tails" of exposure arise from electrons that have been backscattered from the substrate into the resist film but at large distances from their original point of entry (see Fig. 3). These tails result in interproximity effects that limit the minimum separation of lines without proximity compensation. Since the latent image in the resist cannot be observed directly, it is necessary to utilize process simulation to describe the quantitative aspects of electron exposure and proximity effects in latent images. Of course, the development process plays a role in the final observed resist profile, as discussed previously by Kyser *et al.* [16]. Proximity effects are discussed in more detail in Section IV.

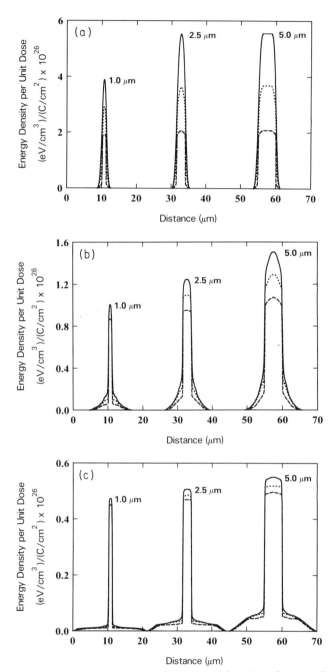

Fig. 6. Absorbed energy density per unit dose as a function of lateral distance in a 1-μm-thick resist film on an infinitely thick silicon substrate at (a) 10, (b) 25, and (c) 50 keV incident energy. Isolated lines with widths of 1.0, 2.5, and 5.0 μm are displayed. The solid, dotted, and dashed curves correspond, respectively, to the bottom, middle, and top 0.05-μm-thick layer in the resist (cell size is 0.05 μm).

D. Spatial Resolution Limits

A major extension of the "single-scattering" Monte Carlo method was made by Murata *et al.* [17] by explicitly including a model for the generation and transport of fast secondary electrons produced in an electron-beam resist by the primary electron exposure. The purpose of that work was to investigate the additional loss of resolution in EBL due to this mechanism. A systematic study was made by Kyser [18,19] on the spatial

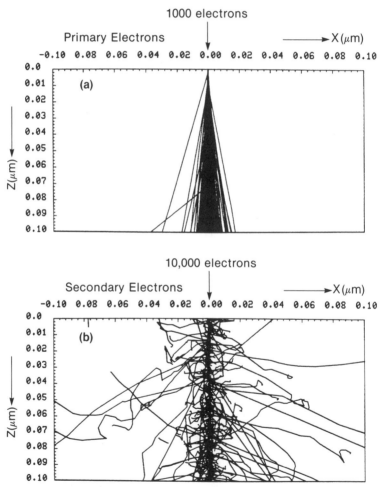

Fig. 7. Point source electron trajectories of (a) 1000 primary electrons and (b) secondary electrons produced by 10,000 primary electrons in a 0.1-μm-thick PMMA foil at an incident energy of 50 keV.

resolution limits in electron-beam "nanolithography" due to such fast secondary electrons with free-standing resist films, including developer effects. An example of the results obtained with this newer hybrid Monte Carlo method is shown in Figs. 7 and 8 for 50-keV irradiation of a 0.1-μm PMMA film without substrate. The primary and secondary electron trajectories are shown in Fig. 7, and the equal energy density contours for specific values of E_v are shown in Fig. 8. Note the additional randomness and noise in the latent image in Fig. 7b due to such a model. This effect may be a real limitation on the fidelity of electron-beam nanolithography. Utilizing a threshold model for development, such calculations appear to

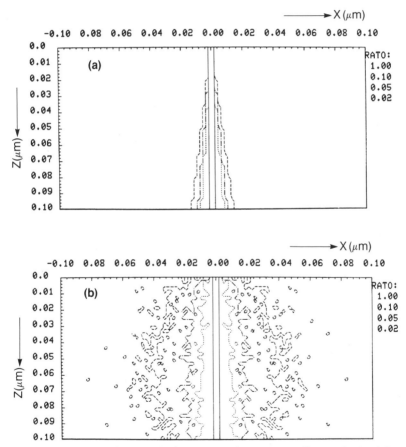

Fig. 8. Equal absorbed energy density contour plots corresponding to Fig. 7: (a) former Monte Carlo model of primary electrons and (b) new hybrid model with fast secondary electrons included.

agree well with the limited amount of experimental data available in the literature. At 100-keV exposure, the model predicts a limiting resolution of ≈ 10 nm in a 0.1-μm-thick PMMA film.

III. ELECTRON RESIST DEVELOPMENT MODELING

The next step in modeling is to include the contributions of the developer solvent to resist profiles. There are two general categories of resist based on their response to absorbed energy. Positive resists become more soluble with increasing absorbed energy, and negative resists become less soluble. For polymeric resists, a change in molecular weight is the mechanism responsible for these changes in solubility. Positive resists chain scission, while negative resists cross-link.

Equal energy density considerations are usually sufficient to predict developed image profiles in negative resists. The solubility of negative resists is usually governed by a critical cross-link density or gel point [20,21]. Areas of the resist with cross-link densities below this critical value remain soluble, whereas areas of the resist with cross-link densities above the critical value are insoluble. As a first approximation, the critical cross-link density is determined by the local absorbed energy density in the resist [22]. Areas of the resist with absorbed energy densities below a certain critical value do not attain the critical cross-link density and remain soluble, whereas areas of the resist with absorbed energy densities equal to or greater than the critical value become insoluble. Therefore, a contour map of the critical absorbed energy density drawn throughout the resist represents the boundary between soluble and insoluble resist. Upon development, this energy contour map is identical with the final developed image profile if swelling is neglected.

Equal energy density considerations alone are not sufficient for positive resists. Positive resists become more soluble with increasing absorbed energy density and generally do not have an analogous minimun critical absorbed energy density below which the resist is insoluble. Even unexposed resist generally remains soluble so that development time and local development rate now become critical to the production of useful images. Developing a positive resist in much the same manner as one would develop a negative resist would result in nothing being left on the wafer, because all the positive resist would be dissolved. Hence, modeling of positive resists must include a development step in addition to the energy density considerations.

The accuracy of the final developed image is dependent on the accuracy of the equations describing the resist response to developer. Modeling of the development process in negative resists is fairly simple, and since there is no time dependence of the development step, the predicted final developed images are reasonably insensitive to small errors in the energy calculations. Positive resists, however, are not as tolerant of calculation error. Small errors in the calculated absorbed energy density are compounded by potentially large errors in the mathematical description of the resist dissolution characteristics. The dissolution theory of positive resist is not as well understood as the absorbed energy density, necessitating the use of empirical equations. In addition, even small errors in development rate are compounded by the time evolution of the development process. These errors are usually manifested in such symptoms as inaccurate predictions of development time and profiles that will not develop the top layer once a line has been opened, leading to predictions of severe undercut that are not observed in real experiments.

The most difficult type of positive resist to model is the so-called nonlinear resist, which exhibits long induction times before developing and is illustrated in Fig. 9. The induction time can be more than 50% of the total development time. The challenge is to find equations that describe the dissolution behavior of the resist without introducing nonphysical behavior such as resist addition (positive slope in the fit to the normalized thickness data is the same as a negative development rate), discontinuities, or infinities. Least-squares fitting of the nonlinear resist dissolution data in Fig. 9 with arbitrary functions such as high-order polynomials results in a

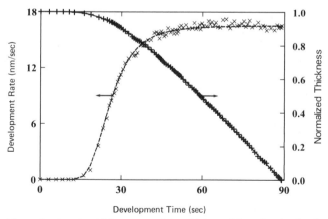

Fig. 9. Normalized resist thickness remaining (right scale) and resist development rate (left scale) as a function of development time for a 1.0-μm thick nonlinear resist film on an infinitely thick silicon substrate blanket exposed with 5.8 μC/cm^2 at 25 keV. Both the experimental data and fitted curves are shown: Eq. (2) to the rate data and Eq. (3) to the thickness remaining data.

reasonable overall fit with superimposed oscillations that lead to the non-physical behavior mentioned above.

Restricting the polynomial least-squares fit to third order or less removes the nonphysical oscillations, but now the resultant fit quality is poor. Least-squares fitting with constraints [23] improves the situation. Imposing the holonomic constraint that the slope of the fit be zero at time zero only suppresses the oscillations at time zero; they remain for all other times. Piecewise polynomial fitting with the holonomic constraint of continuity across the joins is more successful. The usual choice is to require that both the functions and slopes be continuous across the joins. A disadvantage of this approach is increased computational complexity and bookkeeping since each block of time has its own fitting equation and coefficients. A nonholonomic and linear constraint such as requiring the slope of the normalized thickness to be less than or equal to zero removes the problem of nonphysical resist addition but now results in a "stair step" normalized thickness.

The exponential relationship used by Kyser and Pyle [8] and shown in Eq. (1) is an improvement over arbitrary polynomials for highly nonlinear resist, but this type of expression overemphasizes the induction time, leading to predictions of severe undercut, which are not observed in real resists:

$$\text{Rate}(X) = (A + BE^n)(1 - \exp[-\alpha X]) + \beta(E). \qquad (1)$$

One good choice for the rate and thickness equations (see Eib [29]) describing the dissolution behavior of nonlinear resists is to base them on Bose–Einstein approximations to the low temperature (≈ 0 K) heat capacity and internal energy associated with lattice vibrations (phonons) in nonmetallic crystalline insulators [24]. Only three adjustable fitting parameters are needed, so there are no oscillations in the least-squares fits, and the quality of the fit is excellent. The only complication is that nonlinear least-squares fitting (gradient search) techniques must be used since the equations cannot be linearized as with polynomials. However, convergence is rapid and stable as long as care is taken around the point $T = 0$. The development rate is fit with Eq. (2) and the normalized thickness with Eq. (3):

$$\text{Rate}(T) = B_0/\cosh(B_1/T^{B_2}) \qquad (2)$$

and

$$\text{Thickness}(T) = 1.0 - [C_0/(\exp(C_1/T^{C_2}) - 1)], \qquad (3)$$

where B_0, B_1, B_2, C_0, C_1, and C_2 are adjustable parameters and T is time.

Figure 9 shows results obtained for a 1.0-μm-thick nonlinear resist on an infinitely thick silicon substrate blanket (flood) exposed with 5.8 μC/cm² at an accelerating potential of 25 keV. The plus and cross signs are experimental data, and the lines through the points are nonlinear least-squares fits to the data with Eqs. (2) and (3). The data came from the analysis of laser end-point detect traces [25,26]. End-point detect uses a visible wavelength laser to illuminate the resist–wafer sandwich while immersed in developer and a photocell to record reflected standing wave signal intensity patterns. The resulting sinusoidal patterns can be analyzed to produce the resist thickness and development rate as a function of time. The dashed line in Fig. 9 through the cross signs is the nonlinear least-squares fit of Eq. (2) to the rate data (left scale), and the solid line through the plus signs is the nonlinear least-squares fit of Eq. (3) to the thickness data (right scale). Figure 10 is the result of fitting Eq. (3) to blanket exposed nonlinear resist at 0.0, 2.9, 4.3, 5.8, and 7.2 μC/cm², all $\approx$ 1.0-μm-thick on silicon and exposed at 25 keV. It should be noted that these rate and thickness relationships are not exact Bose–Einstein relationships, nor do they share an exact integral–derivative relationship as in low-temperature physics. However, the envelopes of the curves are nearly identical with the Bose–Einstein approximations, and the quality of the least-squares fitting is excellent, especially with the addition of the exponents B_2 and C_2. This suggests that there may be a time and temperature equivalence between resist dissolution and low temperature thermal properties of insulators that is not apparent at this time to the authors. The values of B_2 and C_2 generally are 0.8–1.2. Mathematical simplicity is gained by not insisting

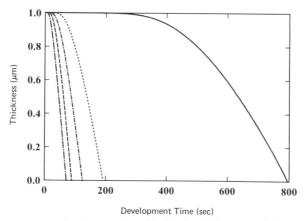

Fig. 10. Least-squares fitted curves [Eq. (3)] of resist thickness remaining as a function of development time for 1.0-μm-thick nonlinear resist on an infinitely thick silicon substrate with exposures of 0.0 (——), 2.9 ($\cdot\ \cdot\ \cdot$), 4.3 ($\cdot-\cdot-$), 5.8 (- - -), and 7.2 ($\cdot-\cdot$) μC/cm² at 25 keV.

on an exact integral–derivative relationship when developing the *master* dissolution rate equation, that is, the rate equation for all absorbed energy densities E and depths X into the resist: Rate(X, E).

The time dependence T is removed by solving Eq. (3) for time and substituting into Eq. (2). Equations (2) and (3) were specifically tailored for this purpose. One need only assume that the exponents B_2 and C_2 are approximately equal to generate Eq. 4:

$$\text{Rate}(X) = \frac{F_0}{[(1 + F_1/X)^{F_2}] + [(1 + F_1/X)^{-F_2}]}, \qquad (4)$$

where F_0, F_1, and F_2 are new combined adjustable least-squares fitting parameters. The best initial guesses to start the nonlinear least-squares fit to the data are $F_0 \approx 2B_0$, $F_1 \approx C_0$, and $F_2 \approx B_1/C_1$. However, this is only part of the X dependence since the absorbed energy density is also a function of depth into the resist (see Fig. 5). The X dependence of the energy is linear (except for thick films at low accelerating potentials) and can be written

$$\text{Energy}(X) = Q(D_0 + D_1 X), \qquad (5)$$

where D_0 and D_1 are, respectively, the intercept and slope of the Monte Carlo blanket exposure absorbed energy density per unit dose [(eV/cm^3)/ (C/cm^2)] graphed in Fig. 5 and Q is the dose (C/cm^2). Combining Eqs. (4) and (5) and allowing for different doses Q results in the *master rate* equation, which is valid for all doses and depths into the resist (F_0 is replaced by the energy density E):

$$\text{Master rate} = \frac{F_4 E^{F_6}}{[(1 + F_1/X)^{F_2}] + [(1 + F_1/X)^{-F_3}]}, \qquad (6)$$

where $E = Q(D_0 + D_1 X)$ from Eq. (5) and one more degree of freedom has been added to the relationship (F_3), which is needed for resist thickness greater than $\approx 1.2\ \mu$m. The fitting parameter values used for all the modeling in this chapter are as follows: $F_4 = 3.378$, $F_6 = 1.248$, $F_1 = 0.1510$, $F_2 = F_3 = 0.7996$, $D_0 = 1.211 \times 10^{26}$, and $D_1 = 0.4814 \times 10^{26}$.

Twenty points equally spaced in time were chosen from each of the five individual (0.0–7.2 μC/cm^2) rate and thickness plots in Fig. 9 and combined using Eq. (6). The results are plotted in Fig. 11. This procedure ensures that each dose Q contributed equally to the master rate least-squares fit. The circles are the data and the lines are the master rate equation fit to the data.

There is one additional refinement to least-squares fitting of the master rate equation. Thinning of the resist in unexposed regions is important for control of defects and later processing such as reactive ion etching (RIE).

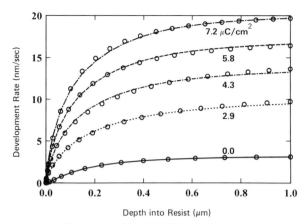

Fig. 11. Least-squares fitted curves [Eq. (7)] of the master rate equation as a function of depth X into nonlinear resist and dose Q [which was transformed into energy E by Eq. (5)], corresponding to Fig. 10. Twenty data points (circles) were taken from each of the five curves in Fig. 10. The line convention is the same as in Fig. 10.

To maximize the fitting accuracy in low-dose regions, one least-squares fits Eq. (6) to the rate *difference* between the zero dose and the rest of the rate data rather than the data itself: rate data$(E > 0)$ − rate data$(E = 0)$. The master rate equation is then transformed back by adding the zero dose rate from Eq. (4):

$$\text{Master rate} = \text{rate}(X,\ E = 0) + \text{rate}(X, E > 0). \tag{7}$$

The residuals of this transformed fit are shown in Fig. 12. The residuals are

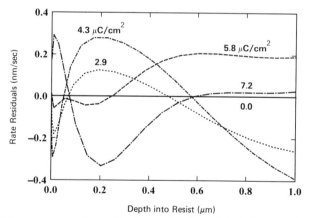

Fig. 12. Master rate equation residuals as a function of depth into nonlinear resist and dose corresponding to Fig. 11. The residuals are defined as the difference between the master rate equation and the data. The line convention is the same as in Fig. 10.

defined as the difference between the fit and the data. This convention is chosen so that a negative residual indicates that the theoretical fit predicts a development rate lower than the experimental data. Equation (7) assures us that the residuals for zero dose (solid line) are zero for all values of X. The accuracy of the master rate fit is about ± 0.3 nm/sec in a range of 10 to 20 nm/sec.

IV. ELECTRON RESIST PROFILE MODELING

A. Comparison of Simulated and Experimental Profiles

Modeling of nonlinear resists can reproduce the experimentally observed biases (Fig. 13a) and sidewall angles (Fig. 13b) with reasonable accuracy and correct predictions of development time ($\pm 3\%$). The dose was 5.0 $\mu C/$ cm^2 at 25 keV in single-layer 1.0-μm-thick nonlinear resist on an infinitely thick silicon substrate. The line openings were measured at the bottom of the resist at the resist–substrate interface. The solid lines and squares are data, and the dashed lines and open circles are modeling predictions. The data are an average of 80 separate wafers. The standard deviation of the average linewidth was approximately ± 0.10 μm, and the standard deviation of the average sidewall angles was approximately $\pm 2°$ based on scanning electron microscope measurements.

B. Proximity Effects and Correction

For much of the following discussion, windage and biases will be considered. *Windage* is defined as the width of the shaped electron-beam spot that must be written in the resist in order for the developed image to be nominal size (zero bias). *Bias* is defined as the difference between the *desired* line dimensions and the *actual* dimension of the developed lines. Maximum bias is defined as the developed image linewidth difference between an n μm (linewidth) by 1.0 μm (space) array and an isolated line where the whole pattern has been windaged to produce zero bias in the isolated feature: maximum bias = $n \times 1$ − isolated. Unless otherwise noted, the line dimensions are measured at the bottom of the features at the resist–substrate interface. Maximum bias reflects intra and interproximity effects, whereas windage reflects only intraproximity effects. The development process point was chosen to be the development time necessary to open a 1.5-μm isolated line to zero bias at 5.0 $\mu C/$cm^2 and 25 keV (2.70 min for 1.0-μm-thick resist, other times for other resist thicknesses).

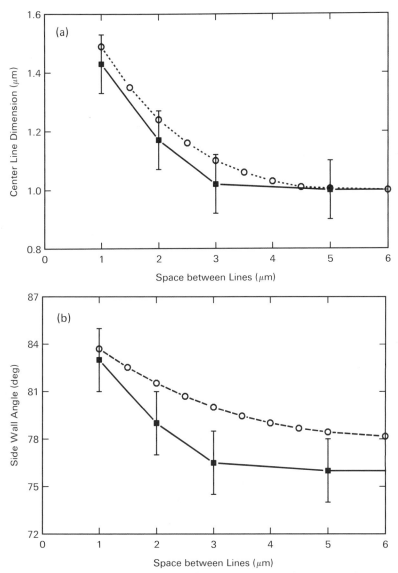

Fig. 13. Comparison of experimental and simulated 1.0-μm-wide array (a) line openings and (b) sidewall angles as a function of line spacing (center to center) in 1.0-μm-thick nonlinear resist on an infinitely thick silicon substrate exposed with 5.0 μC/cm² at 25 keV. The standard deviation of the data is (a) ±0.1 μm and (b) ±2° based on 80 separate wafers measured by scanning electron microscope. The line openings (a) were measured at the bottom of the resist at the resist–substrate interface. ---, simulation; ——, experiment.

The dose necessary for zero bias in the 1.5-μm isolated line at 10 and 50 keV was then adjusted to reproduce the same development time as at 25 keV. Other isolated linewidths were then windaged using these doses and development time until zero bias resulted. Adjustment of the doses rather than the development times was chosen because the developed images were more self-consistent, especially with regard to the side wall angles.

One of the major factors that can limit the general application of EBL to device fabrication is the proximity effect, particularly as device dimensions become smaller. Control of proximity effects becomes crucial as EBL pushes into the submicrometer regime. Fortunately, by its accuracy in predicting images, modeling has proved to be a useful tool for the prediction of proximity effects and for a fundamental understanding of their underlying causes. With understanding comes the possibility of some control. Eight methods for controlling proximity will now be examined with modeling.

1. Electron-Beam Voltage Effects

The first and one of the most important methods for controlling proximity effects is to change the accelerating potential of the electron-beam source. Figures 6a–c are plots of the absorbed energy density per unit dose as a function of position in 1.0-μm-thick resist on an infinitely thick silicon substrate at three accelerating potentials, 10, 25, and 50 keV, for three linewidths, 1.0, 2.5, and 5.0 μm. The absorbed energy density per unit dose has been displayed in the top, middle, and bottom layers in the resist. Note that, as is usual in electron-beam exposures, the bottom layer absorbs the greatest amount of energy and the top layer the least. The edge width of the beam in all cases was 0.10 μm.

At 10 keV (Fig. 6a) there is a large variation in absorbed energy density from top to bottom in the resist. This absorbed energy density gradient saturates for lines ≈ 1.5 μm (not shown) and wider. Also, there is little interproximity effect tailing, as can be seen at 25 keV (Fig. 6b) and especially at 50 keV (Fig. 3c). From an energy consideration alone it would seem that resist exposed at 10 keV would have minimum proximity effects. At 25 and 50 keV the absorbed energy density gradient is considerably reduced, but now the interproximity effect tails begin to dominate. At 25 keV the tails extend ≈ 4 μm out from the edge of the written line and nearly 12 μm at 50 keV.

2. Resist Thickness Effects

The second and probably next most important factor affecting proximity and its control is the thickness of the resist. Figures 14a,b are plots of

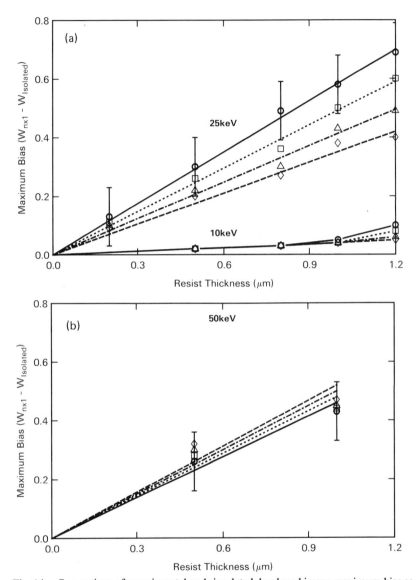

Fig. 14. Comparison of experimental and simulated developed image maximum bias as a function of nonlinear resist thickness for various line widths on an infinitely thick silicon substrate with (a) 1.8 $\mu C/cm^2$ at 10 keV and 5.0 $\mu C/cm^2$ at 25 keV, and (b) 10.17 $\mu C/cm^2$ at 50 keV. Maximum bias is the difference in bias between the center line in an $(n \times 1)$ μm line array and the n μm isolated line, where the isolated line has been windaged to zero bias. The line widths (n) are 1.0 μm (O, ———), 1.5 μm (□, · · ·), 2.5 μm (△, · – · –), and 5.0 μm (◇, – – –). The standard deviation of the data is ±0.1 μm.

TABLE I

Windage and Maximum Bias as a Function of Nonlinear Single-Layer Resist Thickness on an Infinitely Thick Silicon Substrate[a,b]

		\multicolumn{10}{c}{Resist Thickness (μm)}								
		\multicolumn{2}{c}{0.2}	\multicolumn{2}{c}{0.5}	\multicolumn{2}{c}{0.8}	\multicolumn{2}{c}{1.0}	\multicolumn{2}{c}{1.2}
Line	keV	W	MB	W	MB	W	MB	W	MB	W	MB
1.0	10			1.00	0.02	1.05	0.03	1.10	0.05	1.15	0.10
1.5				1.50	0.02	1.50	0.03	1.50	0.04	1.50	0.08
2.5				2.50	0.02	2.50	0.03	2.50	0.04	2.50	0.06
5.0				5.00	0.02	5.00	0.03	5.00	0.04	5.00	0.05
Dose (μC/cm^2)				\multicolumn{2}{c}{1.93}	\multicolumn{2}{c}{1.88}	\multicolumn{2}{c}{1.94}	\multicolumn{2}{c}{1.86}
1.0	25	1.05	0.13	1.20	0.30	1.15	0.49	1.20	0.58	1.15	0.69
1.5		1.50	0.11	1.50	0.26	1.50	0.36	1.50	0.50	1.50	0.60
2.5		2.45	0.10	2.35	0.22	2.35	0.30	2.30	0.43	2.50	0.49
5.0		4.90	0.09	4.75	0.20	4.70	0.27	4.60	0.38	4.55	0.40
Dose (μC/cm^2)		\multicolumn{2}{c}{5.00}	\multicolumn{2}{c}{5.00}	\multicolumn{2}{c}{5.00}	\multicolumn{2}{c}{5.00}	\multicolumn{2}{c}{5.00}
1.0	50			1.05	0.26			1.05	0.43		
1.5				1.50	0.28			1.50	0.44		
2.5				2.45	0.30			2.45	0.45		
5.0				4.85	0.32			4.85	0.47		
Dose (μC/cm^2)				\multicolumn{2}{c}{9.95}			\multicolumn{2}{c}{10.17}		

[a] The maximum bias data is graphed in Figs. 14a,b.

[b] W denotes windage in micrometers; MB denotes maximum bias in micrometers.

maximum bias for three accelerating potentials as a function of single-layer resist film thickness. The windages alone with the resultant biases are tabulated in Table I. The error bars in Fig. 14 represent the usual experimental error of one standard deviation ($\approx \pm 0.10\ \mu$m).

It is apparent from Figs. 14a,b that single-layer resist will always have bias problems unless one can operate with either thin films or at 10 keV. Thin single-layer films usually have pinhole and defect problems. The modeling data extend only to a maximum thickness of 1.2 μm because at the next thickness (1.4–1.6 μm) large features such as 5 × 1 μm arrays cannot maintain shape, the spacing between the lines is destroyed by the interproximity effect.

The dependence of bias on thickness is linear and seems to extrapolate approximately to zero. The bias dependence of 25 and 50 keV single-layer films is also nearly equal. That is to be expected since the modeling was based on infinite line arrays, and at accelerating potentials greater than ≈ 25 keV, the total electron scattering is dominated by the substrate. A very rough way to measure the number of scattering electrons responsible for proximity effects is to sum the number of electrons that backscatter from the film–substrate target into the vacuum above the resist. At 25 keV and 1.0-μm resist on silicon, 13.5% of the electrons backscatter from the top surface of the resist, whereas at 50 keV the percentage is 15.5.

At 10 and 25 keV, the narrowest line (1.0 μm) exhibits the greatest maximum bias, whereas at 50 keV, the widest line (5.0 μm) exhibits the greatest maximum bias. This reversal of the maximum bias dependence on linewidth at 50 keV is due to the windages added to the patterns. Without windage, the patterns do not reverse. The closer grouping together of the various linewidths at 50 keV compared with 25 keV is due to the longer scattering range of the electrons where intraproximity effects are even less important than at 25 keV. With higher accelerating potentials, the individual lines and their spacing are less important than the average writing density in the whole pattern. For example, averaged over many lines, 1×1 μm arrays result in the same total energy being delivered to the resist as 1.5×1.5 μm or 5×5 μm arrays. Similar absorbed energy density roughly results in similar developed image biases.

3. Substrate and Multilayer Resist Effects

Four additional methods for reducing proximity at 10, 25, and 50 keV in 0.5–1.0-μm-thick resist are tabulated in Tables II–IV, respectively, and graphed in Figs. 15a–c. As might be expected from the very small proxim-

TABLE II

Windage and Maximum Bias in Nonlinear Resist at 10 keV for
Five Resist–Substrate Configurations[a,b,c]

| | | Type of Image | | | | | | | | |
| | | SLR | | SLR(Au) | | MLR | | TS | | TS,MLR | |
Line	Thickness	W	MB	W	MB	W	MB	W	MB	W	MB
1.0	1.0	1.10	0.05	1.15	0.05	0.85	0.01	0.85	0.01	0.85	0.01
1.5	and	1.50	0.04	1.50	0.03	1.35	0.01	1.35	0.01	1.35	0.01
2.5	0.5	2.50	0.04	2.45	0.03	2.35	0.01	2.35	0.01	2.35	0.01
5.0		5.00	0.03	4.95	0.02	4.85	0.01	4.85	0.01	4.85	0.01
Dose (μC/cm^2)		1.94		1.78		1.94		3.87		3.87	

[a] The five configurations are as follows: SLR is single-layer resist on infinitely thick silicon, SLR(Au) is single-layer resist on infinitely thick gold, MLR is multilayer resist with a total thickness of 3.0 μm on infinitely thick silicon, TS is top surface imaging in single-layer resist on infinitely thick silicon, and TS,MLR is top surface imaging in the same multilayer stack as MLR.

[b] W denotes windage in micrometers; MB denotes maximum bias in micrometers.

[c] The maximum bias data is graphed in Fig. 15a.

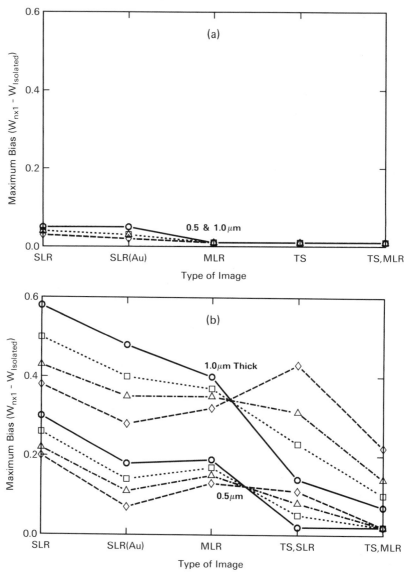

Fig. 15. Simulation of developed image maximum bias as a function of image type for various line sizes and nonlinear resist thicknesses at (a) 10, (b) 25, and (c) 50 keV. SLR is single-layer resist on infinitely thick silicon. SLR(Au) is single-layer resist on infinitely thick gold. MLR is multilayer resist with a total thickness of 3.0-μm on infinitely thick silicon. TS is top surface imaging in single-layer resist on infinitely thick silicon. TS,MLR is top surface imaging in the same multilayer stack as MLR. The symbol and line convention are the same as in Fig. 14.

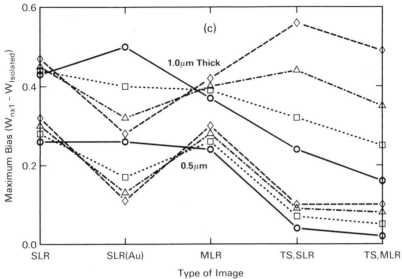

Fig. 15 *(Continued)*

TABLE III

Windage and Maximum Bias in Nonlinear Resist at 25 keV for
Five Resist–Substrate Configurations[a,b,c]

		Type of Image								
		SLR		SLR(Au)		MLR		TS		TS,MLR	
Line	Thickness	W	MB	W	MB	W	MB	W	MB	W	MB
---	---	---	---	---	---	---	---	---	---	---	---
1.0	1.0	1.20	0.58	1.25	0.48	1.10	0.40	0.80	0.14	0.85	0.07
1.5		1.50	0.50	1.50	0.40	1.50	0.37	1.25	0.23	1.30	0.10
2.5		2.30	0.43	2.30	0.35	2.40	0.35	2.20	0.31	2.30	0.14
5.0		4.60	0.38	4.70	0.28	4.75	0.32	4.55	0.43	4.70	0.22
Dose (μC/cm²)		5.00		3.41		5.44		10.00		10.88	
1.0	0.5	1.20	0.30	1.15	0.18	1.05	0.19	0.95	0.02	0.95	0.02
1.5		1.50	0.26	1.50	0.14	1.50	0.17	1.45	0.05	1.45	0.02
2.5		2.35	0.22	2.40	0.11	2.40	0.15	2.40	0.08	2.45	0.02
5.0		4.75	0.20	4.90	0.07	4.85	0.13	4.90	0.11	4.95	0.02
Dose (μC/cm²)		5.00		3.22		5.39		10.00		10.78	

[a] The five configurations are as follows: SLR is single-layer resist on infinitely thick silicon, SLR(Au) is single-layer resist on infinitely thick gold, MLR is multilayer resist with a total thickness of 3.0 μm on infinitely thick silicon, TS is top surface imaging in single-layer resist on infinitely thick silicon, and TS,MLR is top surface imaging in the same multilayer stack as MLR.

[b] W denotes windage in micrometers; MB denotes maximum bias in micrometers.

[c] The maximum bias data is graphed in Fig. 15b.

TABLE IV

Windage and Maximum Bias in Linear Resist at 50 keV for
Five Resist–Substrate Configurations[a,b,c]

		Type of Image								
		SLR		SLR(Au)		MLR		TS		TS,MLR	
Line	Thickness	W	MB	W	MB	W	MB	W	MB	W	MB
1.0	1.0	1.05	0.43	1.20	0.50	1.05	0.37	0.85	0.24	0.85	0.16
1.5		1.50	0.44	1.50	0.40	1.50	0.39	1.30	0.32	1.33	0.25
2.5		2.45	0.45	2.30	0.32	2.45	0.40	2.30	0.44	2.30	0.35
5.0		4.85	0.47	4.70	0.28	4.85	0.42	4.70	0.56	4.70	0.49
Dose (μC/cm^2)		10.17		6.95		10.24		20.34		20.48	
1.0	0.5	1.05	0.26	1.20	0.26	1.05	0.24	0.95	0.04	0.95	0.02
1.5		1.50	0.28	1.50	0.17	1.50	0.26	1.45	0.07	1.45	0.05
2.5		2.45	0.30	2.40	0.13	2.45	0.28	2.45	0.09	2.45	0.08
5.0		4.85	0.32	4.85	0.11	4.85	0.30	4.90	0.10	4.90	0.10
Dose (μC/cm^2)		10.17		6.50		10.24		20.34		20.48	

[a] The five configurations are as follows: SLR is single-layer resist on infinitely thick silicon, SLR(Au) is single-layer resist on infinitely thick gold, MLR is multilayer resist with a total thickness of 3.0 μm on infinitely thick silicon, TS is top surface imaging in single-layer resist on infinitely thick silicon, and TS,MLR is top surface imaging in the same multilayer stack as MLR.

[b] W denotes windage in micrometers; MB denotes maximum bias in micrometers.

[c] The maximum bias data is graphed in Fig. 15c.

ity effects observed in single-layer resist at 10 keV, additional resist–substrate configurations are unnecessary at 10 keV but have been included for the sake of completeness.

The third method of reducing proximity effects is to add an electron scattering, high-atomic-number layer directly under the resist layer, such as infinitely thick gold (at 25 keV, this is ≈ 0.4 μm). The gold layer narrows the lateral spread of the backscattered electrons (interproximity is reduced) and greatly increases their number as can be seen in Figs. 3 and 4. Intra-proximity effects are actually increased at 25 and 50 keV, resulting in an average increase in resist sensitivity of $\approx \frac{1}{3}$. The required doses decrease from 5.0 to 3.44 μC/cm^2 at 25 keV, and from 10.17 to 6.95 μC/cm^2 at 50 keV. The percentage of electrons backscattered into the vacuum above 1.0-μm-thick single-layer resist increases from 13.5% on silicon to 45.2% on gold at 25 keV, and from 15.5% on silicon to 46.4% on gold at 50 keV. Another way to characterize this increased sensitivity is to look at development times at constant dose: times to develop blanket exposed regions are reduced by half. Simulation predicts a decrease in the blanket exposure development time from 1.79 to 0.94 min, while experimentally the decrease is from 1.8 to 0.90 min (all at 5.0 μC/cm^2 and 25 keV).

The fourth method of reducing proximity effects is to add an electron-transmitting, low-atomic-number layer directly under the resist layer, such as 3 μm of inert polymer. This technique reduces both intra- and inter-proximity effects by moving the imaging layer away from the silicon substrate, thereby reducing the probability of atomic interaction with the backscattering electrons from the substrate.

The resist–substrate sandwich in Tables II–IV and Fig. 15 consisted of, respectively, 1.0 or 0.5 μm resist on 2.0–2.5 μm inert polymer for a constant total thickness of 3.0 μm. It was assumed for the Monte Carlo calculations that the composition of the inert polymer is the same as the resist (60 wt % carbon, 32 wt % oxygen, and 8 wt % hydrogen) on infinitely thick silicon. The dose necessary for the 1.5-μm isolated line to open to zero bias at 2.70 min increased from 5.0 to 5.44 μC/cm^2 in 1.0-μm resist, and to 5.39 μC/cm^2 in 0.5-μm resist. Thicker inert polymer films were not considered because the reduction of proximity effects was minimal. Also, beam registration and sample charging become a problem at total thickness much greater than ≈ 3–4 μm. The reversal of the biases at 50 keV is again due to the windages built into the pattern. The net effect of using multilayer techniques is to reduce proximity slightly compared with single-layer resist and group the biases closer together. Experimentally, the reversal is not distinguishable because experimental error is usually at least ± 0.10 μm (one standard deviation). As with all multilayer techniques, the resist image would then have to be transferred into the underlying inert polymer by some method such as RIE. RIE usually requires resist hardening by some technique such as backing, ultraviolet exposure, chemical modification, or plasma hardening. These additional processing steps along with the RIE transfer step introduce additional biases in the final image.

The fifth method of reducing proximity effects is to adopt top surface imaging techniques [27]. Top surface imaging in single-layer resist is listed as image type TS in Tables II–IV, and multilayer resist as image type TS,MLR. The technique consists of choosing a process point such that developed image sidewall angles are greater than 90°. This insures that the top of the image determines the transferred line size rather than the bottom of the resist. The top of the resist absorbs less energy (see Fig. 5), and develops more slowly than the bottom of the resist and is therefore less sensitive to fluctuation in either the absorbed energy density (intra- and interproximity effects) or development time.

Top surface imaging requires higher doses to produce sidewall angles greater than 90°. The actual dose required to produce undercut images was ≈ 8.5–9 μC/cm^2 as 25 keV, so an exact double was close enough. The development time was chosen to produce 90° walls in the 1.0-μm isolated line rather than the 1.5-μm isolated line. The time difference was not significant because the top surface images are fairly insensitive to overdevelopment. Windage had a much greater effect than overdevelopment. The least amount of bias occurred when top surface imaging was combined with multilayer techniques in image type TS,MLR. However, top surface imaging has the same additional processing requirements as multilayer techniques. The resist usually must be hardened before image transfer into the substrate.

In thin films, the reduction in proximity effects from employing top surface imaging in multilayer resist is great enough to obviate the need for additional techniques such as dose correction. In fact, this is the construction usually chosen for negative resists. A very thin negative imaging layer approximately $0.1-0.3$-μm thick is constructed as a multilayer, where cross-linking induced by the electron beam produces a partial top surface image part way down in the resist. In addition, the cross-linked resist can sometimes act as its own RIE barrier without additional hardening. The major key to reducing biases, however, is the thinness of the imaging layer rather than the use of multilayer techniques.

The sixth method of reducing proximity effects is to switch to a linearly developing resist such as PMMA. The more sensitive nonlinear resists generally consist of two components: an inert (to electron irradiation) soluble resin, and an electron sensitive dissolution inhibitor. Irradiation destroys the dissolution inhibitor, thereby rendering the exposed resist more soluble than the unexposed. The two-component nonlinear systems are generally more sensitive than the one-component linear systems but will dose saturate. Once all the dissolution inhibitor is destroyed, further irradiation does not produce greater solubility as is the case with linear PMMA. In fact, overexposure of nonlinear resist systems can result in total loss of images. Interproximity effects completely expose the areas of the resist not directly addressed by the electron beam. Linear resists may therefore adopt the optical photographic technique of overexposing and underdeveloping at the expense of throughput. Linear and nonlinear resists are compared in Tables V and VI at 25 and 50 keV, respectively, and graphed in Fig. 16a,b.

The seventh method of reducing proximity effects is to use thin substrates. Table VII and Fig. 17 illustrate the advantages to using finite-thickness substrates at 25 and 50 keV. This technique is used for the fabrication of x-ray masks in which silicon membranes are employed. The zero-thickness silicon membrane has been included to characterize the resistant self-contribution to intraproximity effects. The self-contribution in 1.0-μm-thick nonlinear single-layer resist is 0.02 μm at 25 and 50 keV. Acceptable biases of 0.1 μm or less in single-layer resist necessitate the use of membranes with maximum thicknesses of 0.5 μm at 25 keV, and

TABLE V

Windage and Maximum Bias in Linear and Nonlinear Resist at 25 keV for Five Resist–Substrate Configurations[a,b,c]

	Type of Image									
	SLR		SLR(Au)		MLR		TS		TS,MLR	
Resist thickness	W	MB	W	MB	W	MB	W	MB	W	MB
1.0										
Nonlinear	1.20	0.58	1.25	0.46	1.10	0.40	0.80	0.14	0.85	0.07
Dose (μC/cm²)	5.00		3.41		5.44		10.00		10.88	
Linear	1.20	0.43	1.25	0.32	1.10	0.28	0.80	0.18	0.85	0.09
Dose (μC/cm²)	10.00		6.82		10.88		20.00		21.76	
0.5										
Nonlinear	1.20	0.30	1.15	0.18	1.05	0.19	0.95	0.02	0.95	0.02
Dose (μC/cm²)	5.00		3.22		5.39		10.00		10.78	
Linear	1.20	0.17	1.15	0.09	1.05	0.12	0.80	0.07	0.85	0.05
Dose (μC/cm²)	10.00		6.82		10.88		20.00		21.76	

[a] The five configurations are as follows: SLR is single-layer resist on infinitely thick silicon, SLR(Au) is single-layer resist on infinitely thick gold, MLR is multilayer resist with a total thickness of 3.0 μm on infinitely thick silicon, TS is top surface imaging in single-layer resist on infinitely thick silicon, and TS,MLR is top surface imaging in the same multilayer stack as MLR.

[b] W denotes windage in micrometers; MB denotes maximum bias in micrometers.

[c] The maximum bias data is graphed in Fig. 16a.

TABLE VI

Windage and Maximum Bias in Linear and Nonlinear resist at 50 keV for Five Resist–Substrate Configurations

	Type of Image									
	SLR		SLR(Au)		MLR		TS		TS,MLR	
Resist thickness	W	MB	W	MB	W	MB	W	MB	W	MB
1.0										
Nonlinear	1.05	0.43	1.20	0.50	1.05	0.37	0.85	0.22	0.85	0.14
Dose (μC/cm²)	10.17		6.95		10.24		20.34		20.48	
Linear	1.05	0.29	1.20	0.19	1.05	0.25	0.85	0.12	0.85	0.11
Dose (μC/cm²)	20.34		13.90		20.48		40.68		40.96	
0.5										
Nonlinear	1.05	0.26	1.20	0.26	1.05	0.23	0.95	0.04	0.95	0.02
Dose (μC/cm²)	5.00		3.22		5.39		10.00		10.78	
Linear	1.05	0.13	1.20	0.10	1.05	0.12	0.95	0.07	0.95	0.05
Dose (μ/cm²)	10.00		6.44		10.78		20.00		21.56	

[a] The five configurations are as follows: SLR is single-layer resist on inifinitely thick silicon, SLR(Au) is single-layer resist on infinitely thick gold, MLR is multilayer resist with a total thickness of 3.0 μm on infinitely thick silicon, TS is top surface imaging in single-layer resist on infinitely thick silicon, and TS,MLR is top surface imaging in the same multilayer stack as MLR.

[b] W denotes windage in micrometers; MB denotes maximum bias in micrometers.

[c] The maximum bias data is graphed in Fig. 16b.

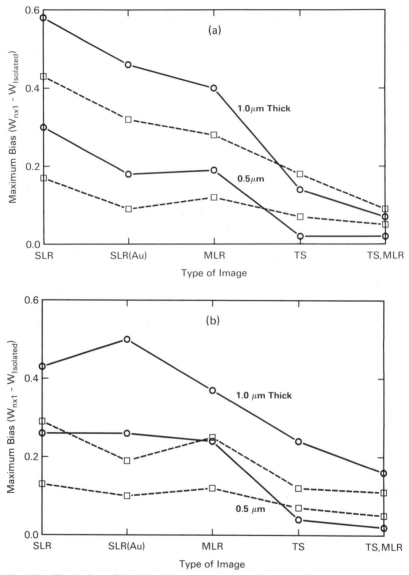

Fig. 16. Simulation of developed image maximum bias as a function image type for 1.0-μm lines in various thicknesses of linear and nonlinear resist at (a) 25 and (b) 50 keV. Behavior at 10 keV is the same as Fig. 15a. The image types are the same as in Fig. 15. ———, Nonlinear resist; – – –, linear resist.

TABLE VII

Windage and Maximum Bias in 1.0-μm-thick Nonlinear Resist at 25 and 50-keV on Finite Thickness Silicon Membranes

		Silicon substrate thickness (μm)								
		0.0		0.5		1.0		1.5		∞	
Line	keV	W	MB	W	MB	W	MB	W	MB	W	MB
1.0	25	1.00	0.02	1.05	0.16	1.10	0.38	1.15	0.56	1.20	0.58
1.5		1.50	0.02	1.50	0.15	1.50	0.32	1.50	0.48	1.50	0.50
2.5		2.50	0.02	2.45	0.14	2.40	0.26	2.35	0.41	2.30	0.43
5.0		5.00	0.02	4.85	0.13	4.75	0.19	4.65	0.33	4.60	0.38
Dose (μC/cm^2)		5.84		5.62		5.39		5.13		5.00	

		0.0		1.0		3.0		5.0		∞	
Line	keV										
1.0	50	1.10	0.02	1.00	0.03	1.00	0.12	1.05	0.26	1.05	0.43
1.5		1.50	0.02	1.50	0.04	1.50	0.15	1.50	0.28	1.50	0.44
2.5		2.50	0.02	2.50	0.05	2.50	0.16	2.45	0.30	2.45	0.45
5.0		5.00	0.02	5.00	0.05	5.00	0.18	4.85	0.32	4.85	0.47
Dose (μC/cm^2)		10.70		10.60		10.48		10.30		10.17	

[a] W denotes windage in micrometers; MB denotes maximum bias in micrometers.
[b] The maximum bias data is graphed in Fig. 17.

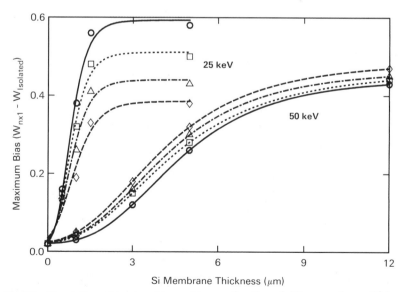

Fig. 17. Simulation of developed image bias as a function of silicon membrane thickness for 1.0-μm-thick nonlinear resist exposed at 25 and 50 keV. The symbol and line convention are the same as in Fig. 14.

3.0 μm at 50 keV. Membranes 0.5-μm thick are very fragile. One would prefer to operate either below 10 keV or above 50 keV.

4. Exposure Modulation Effects

The eighth and perhaps most successful approach for controlling proximity effects at all beam voltages is a full dose correction. Pattern windages are simplified: windage based on line size may be replaced with one general windage applied to all line sizes. Two approaches to calculating the proximity-corrected doses have been made and the results compared.

The first was an energy calculation that equalized the absorbed energy density at the center of the exposed line shapes, and the second was a complete energy–develop iteration that adjusted the doses until all images developed to zero bias. The first method is an energy-only consideration; the second includes the contributions of the developer to final images. The calculations were made for single-layer 1.0-μm resist on infinitely thick silicon at 10, 25, and 50 keV. The doses were normalized to produce a development time of 2.70 min for zero bias in the 1.5-μm isolated line.

Figures 18a–c are graphs of the proximity corrected patterns by the full energy–develop correction method. Only interior lines in an infinite line array are displayed (infinite from a proximity point of view). Figures

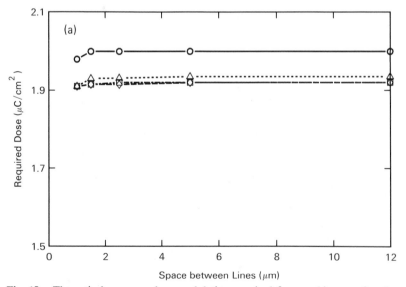

Fig. 18. Theoretical exposure dose modulation required for zero bias as a function of infinite array line spacing in 1.0-μm-thick nonlinear resist on an infinitely thick silicon substrate exposed at (a) 10, (b) 25, and (c) 50 keV. The simulation includes development. The symbol and line convention are the same as in Fig. 14.

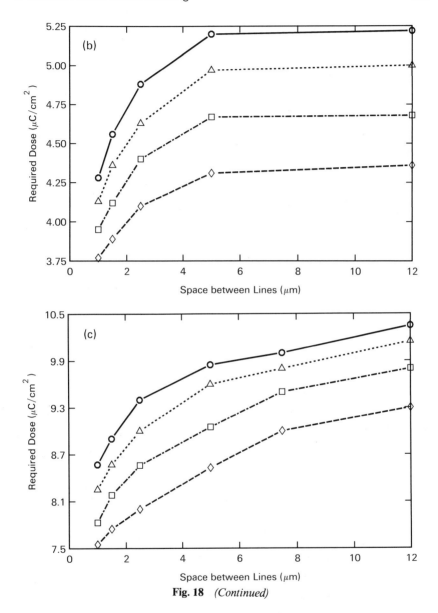

Fig. 18 *(Continued)*

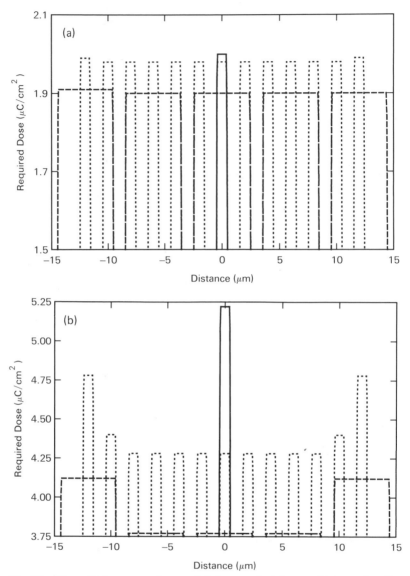

Fig. 19. Theoretical exposure does modulation required for zero bias as a function of finite array line spacing in 1.0-μm-thick nonlinear resist on an infinitely thick silicon substrate exposed at (a) 10, (b) 25, and (c) 50 keV. The array dimensions are 1.0-μm isolated line (——), 1.0-μm line by 1.0-μm space ($\cdot\cdot\cdot$), and 5.0-μm line by 1.0-μm space (– – –). The simulation includes development.

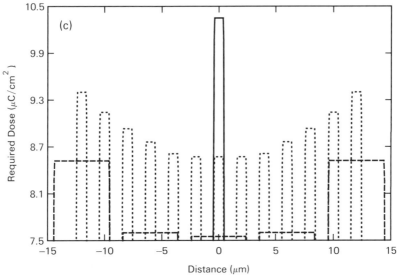

Fig. 19 *(Continued)*

19a–c display the lateral dose variation of proximity corrected doses in an *finite* array for three line sizes: $1 \times \infty$ μm (isolated), 1×1 μm, and 5×1 μm. The solid line is the 1.0 μm isolated line, the dotted lines are 1×1 μm, and the dashed lines are 5×1 μm. These three arrays were chosen to illustrate the highest average dose correction [$(1 \times \infty)$ μm array], the lowest average dose correction [(5×1) μm array], and the greatest spatial dose variation [(1×1) (μm. array].

At 10 keV there is little dose variation between either the various interior line–space combinations in an infinite array or spatial variation in finite arrays. This prediction is supported experimentally. Little if any dose correction is needed when one is operating at 10 keV. The prediction, however, is quite different if only an energy correction is made without including the contributions of the developer, where the large top-to-bottom dose gradient in the resist dominates the calculation. The differences in calculated dose between the energy-only and energy–develop iteration are plotted in Figs. 20a–c. Hence at 10 keV, the prediction is that no proximity correction is needed, but, if done by an energy-only correction scheme, will result in the wrong answer.

At 25 keV, however, proximity correction is definitely required. The agreement between the two calculation methods is quite good, so that either method could be used for proximity correction. The average difference is 0.1 μC/cm² in a total average dose of 5.0 μC/cm². One would, of

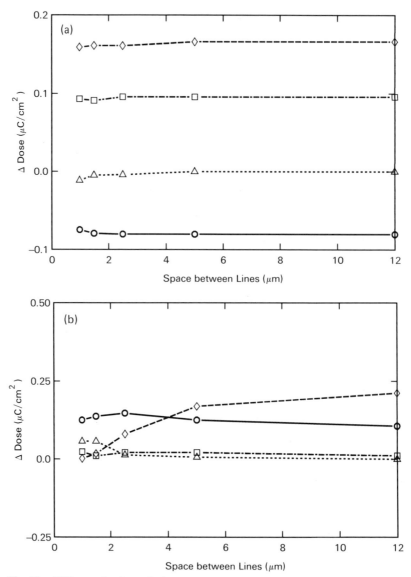

Fig. 20. Difference in theoretical exposure dose modulation required for zero bias as a function of infinite array line spacing in 1.0-μm-thick nonlinear resist on an infinitely thick silicon substrate exposed at (a) 10, (b) 25, and (c) 50 keV. The dose difference is between a complete energy/develop simulation and an energy only simulation. The symbol and line convention are the same as in Fig. 14.

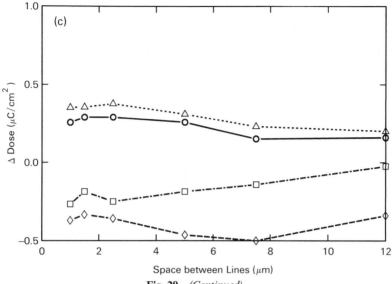

Fig. 20 *(Continued)*

course, choose the easier method of calculation, which is the energy-only approach. The maximum electron scattering range is such that only nearest neighbors need be considered even in the 1×1 μm array. Experimentally, the average difference between calculation by either method and experiment was ≈ 0.3 μC/cm^2 (6%) with a maximum difference of 0.4 μC/cm^2. Hence, at 25 keV an energy-only proximity correction will be sufficient.

The difference in methods becomes large again at 50 keV, although not as large as at 10 keV. The developer contributions are important again and the large ≈ 12-μm scattering range of the electrons requires considerations of fifth and sixth neighbors in the 1×1 μm array. The net result is that energy considerations alone are insufficient for proximity correction at 50 keV. One must use a full energy–develop correction.

It should be pointed out that not all proximity correction schemes are entirely energy-only calculations. Many schemes use analytical functions to calculate the absorbed energy density in the resist such as a double Gaussian approximation for the forward scattering in the resist and backward scattering from the substrate [22,28]. This model attempts to include developer contributions to the developed images by empirically adjusting three parameters that affect the total absorbed energy until the best fit is found with experimental data: the Gaussian full width at half maximum (FWHM) of the forward-scattered electrons, the Gaussian FWHM of the backward-scattered electrons from the substrate, and the percentage each Gaussian contributes to the total energy. This scheme works fairly well at 25 keV but difficulties are encountered at 50 keV because three adjustable parameters are too few to characterize both the energy and increased developer contributions to line profiles in nonlinear resist systems [30].

V. CONCLUSIONS

Theoretical modeling of the electron-beam resist process is a valuable aid to experimental design and interpretation. General modeling of electron scattering and energy deposition in resist-film – silicon – substrate targets has been accomplished with Monte Carlo calculations. Such Monte Carlo calculations can treat a wide variety of target configurations, including foil substrates, without additional approximations. The modeling includes experimental parameters such as primary beam voltage, resist film and substrate thickness, resist and substrate composition, exposure pattern, and developer effects. Both intraproximity and interproximity effects are automatically included in the simulation.

The spatial distribution of energy deposition within the resist film or latent image is transformed into a solubility rate image in order to calculate and display the time evolution of the developed image. The transformation is accomplished by utilizing a master rate equation with appropriate parameters. The quantitative values of the parameters are deduced by sophisticated and extremely accurate curve fitting of an appropriate algorithm to experimental data on development rate as a function of development time for a plane exposure. This technique is particularly useful for a class of nonlinear resist materials that exhibit a time delay or induction time in the onset of a rapid development rate. Such resist materials can provide technological improvements in exposure sensitivity and profile shape, compared with linear resist materials such as PMMA.

The accuracy of the theoretical simulation has been verified by comparison with experimental results over a range of line shapes for 1.0-μm thin resist on silicon substrates at 25-kev exposure. The accuracy of the simulation gives confidence in the methods utilized, and allows application to other materials and experimental conditions with a high degree of confidence.

Modeling has been applied to the investigation of the causes and control of proximity effects in EBL. Intraproximity effects are corrected by pattern windages. Interproximity correction schemes can be grouped into five main categories in approximate order of their importance.

(1) *Dose correction.* Energy considerations alone are sufficient for calculating the correct doses at 25 keV. The developer contributions to the final images are not significant, and only nearest neighbors need be considered. At 50 keV, however, dose correction is much more difficult. The calculation must include developer effects and must examine features up to 10 μm distant since nearest neighbors are no longer sufficient. The major disadvantage of dose correction is the additional computer usage required to process the patterns. At 10 keV there is no need for proximity correction.

(2) *Accelerating potential.* Lower accelerating potentials result in lower interproximity effects. The lowest *practical* accelerating potential is usually set by electron optics, beam registration, and the requirement that the bottom of the resist be exposed. The resist thickness sets an absolute lower limit on the accelerating potential. For example, the total electron penetration depth at 5 keV is 0.5 μm. This sets the maximum useful resist thickness to ≈ 0.3 μm at 5 keV.

Beam registration also raises the lowest practical accelerating potential. Registration is a process whereby pre-existing features (usually high Z metal lines in the substrate) are used to position the electron beam for the next set of exposures. This may be accomplished by monitoring for backscattered electrons as the beam is dithered over the features. The short mean free path of low-energy electrons, which is responsible for reducing interproximity effects nearly to zero, also results in the absorption of these backscattering registration electrons by the resist. One may either operate without registration, add lithographic steps to uncover the registration lines, use optical registration, or operate at higher energy.

One last consideration for the lowest practical accelerating potential is susceptibility to stray magnetic or electric fields such as those produced by resist charging.

(3) *Resist thickness.* Thinner resist films result in lower intraproximity and interproximity effects. The minimum *practical* film thickness is set by considerations of pinholes (defects), step coverage, and later processing steps such as RIE resistance. One generally uses multilayer techniques to alleviate the step coverage and some of the pinhole and RIE limitations.

(4) *Substrate effects.* There is some advantage either to using high atomic number Z metals directly under the resist or to moving the imaging layer away from the substrate backscattered electrons by means of a low-scattering polymeric layer (multilayer techniques). The first method takes advantage of the reduced lateral spread of the backscattering electrons to reduce interproximity effects and their increased number to improve sensitivity. The latter method attempts to diffuse the backscattered electrons over wide areas to reduce their contribution to the total exposure. Other

techniques such as top surface imaging take advantage of nonlinear resist dissolution characteristics and the lower absorbed energy density at the top of the resist layer as compared with the bottom.

(5) *Resist choice.* Linear and nonlinear resists exhibit the same intra-proximity effects. The lower contrast linear resist generally exhibits smaller interproximity effects than the higher contrast nonlinear resist. However, nonlinear resist by its very nature does not thin appreciably as compared with linear resist (this is important for RIE resistance), it exhibits comparable proximity effects in the more complicated structures such as multi-layer techniques with top surface imaging, it dose corrects the same as the linear resist at 25 keV, and it is more sensitive, reducing the required exposure dose from $10-15$ $\mu C/cm^2$ to $2-5$ $\mu C/cm^2$.

No one method is usually sufficient to control proximity effects. Many or all of the methods are combined to reduce proximity effects to the lowest possible degree. A typical example would be a thin imaging layer on a high-Z metal layer in a multilayer structure. The exposure dose would be large enough to take advantage of top surface imaging, and the accelerating potential would be set as low as possible. Dose correction would then be employed to compensate for all remaining proximity effects.

In summary, the rapid investigation of EBL processes via simulation is an important component of the art and science of the technology and will become even more important in the era of VLSI. As EBL becomes more mature and pattern dimensions become ever smaller, process simulation will become a necessity. Simulation will certainly be used now and in the future as a powerful predictive tool in the design of superior theoretical resists for the next generation of EBL materials. Perhaps simulation will even help guide the synthesis of these new resist materials.

ACKNOWLEDGMENTS

The authors thank D. Myers of IBM, East Fishkill, for the dissolution curves and L. Gregor for much of the experimental image data presented in this chapter. We thank the management of IBM and Philips for the support we have received in the course of this work. The work by D. Kyser was done as a member of the research staff at the IBM Research Lab, San Jose, California.

REFERENCES

1. P. Thornton, *Adv. Electron. Electron Phys.* **48,** 271 (1979).
2. P. Thornton, *Adv. Electron. Electron Phys.* **54,** 69 (1980).

3. E. Munro, *Adv. Electron. Electron Phys., Suppl.* **13B**, 73 (1980).
4. R. Hawryluk, A. Hawryluk, and H. Smith, *J. Appl. Phys.* **45**, 2551 (1979).
5. J. Greeneich and T. Van Duzer, *IEEE Trans. Electron Devices* **ED-21**, 286 (1974).
6. D. Kyser and K. Murata, *Proc. Int. Conf. Electron Ion Beam Sci. Technol., 6th* p. 205. Electrochem. Soc., Princeton, New Jersey, 1974.
7. A. Neureuther, D. Kyser, and C. Ting, *IEEE Trans. Electron Devices* **ED-26**, 686 (1979).
8. D. Kyser and R. Pyle, *IBM J. Res. Dev.* **24**, 426 (1980).
9. S. Chang, D. Kyser, and C. Ting, *IEEE Trans. Electron Devices* **ED-28**, 1295 (1980).
10. R. Hawryluk, *J. Vac. Sci. Technol.* **19**, 1 (1981).
11. F. Jones and J. Paraszczak, *IEEE Trans. Electron Devices* **ED-28**, 1544 (1981).
12. N. Eib and F. Jones, *J. Vac. Sci. Technol., B* **1**, 1372 (1983).
13. W. Moreau, *Opt. Eng.* **22**, 181 (1983).
14. D. Kyser, *Scanning Electron Microsc.* **1**, 47 (1981).
15. K. Murata, *in* "Electron Beam Interactions with Solids," p. 311. SEM, Chicago, Illinois, 1984.
16. D. Kyser, D. Schreiber, and C. Ting, *Proc. Int. Conf. Electron Ion Beam Sci. Technol., 9th* p. 255. Electrochem. Soc., Princeton, New Jersey, 1980.
17. K. Murata, D. Kyser, and C. Ting, *J. Appl. Phys.* **52**, 4396 (1981).
18. D. Kyser, *J. Vac. Sci. Technol., B* **1**, 1391 (1983).
19. D. Kyser, *in* "Electron Beam Interactions with Solids," p. 331. SEM, Chicago, Illinois, 1984.
20. N. Atoda and H. Kawakatsu, *J. Electrochem. Soc.* **123**, 1519 (1976).
21. A. Reiser and E. Pitts, *J. Photogr. Sci.* **29**, 187 (1981).
22. J. Paraszczak, D. Kern, M. Hatzakis, J. Bucchignano, E. Arthur, and M. Rosenfield, *J. Vac. Sci. Technol., B* **1**, 1372 (1983).
23. S. L. Meyer, "Data Analysis for Scientists and Engineers," Chap. 33. Wiley, New York, 1975.
24. C. Kittel, "Introduction to Solid State Physics," 3rd Ed., Chap. 6. Wiley, New York, 1967.
25. K. Konnerth and F. Dill, *IEEE Trans. Electron Devices* **ED-22**, 452 (1975).
26. L. Thompson, C. Willson, and M. Bowden, "Introduction to Microlithography," ACS Symposium Series No. 219, p. 102. Am. Chem. Soc., Washington, D.C., 1983.
27. S. Gillespie, *Proc. Microcircuit Eng. '82* p. 16. Sitecmo Dieppe, Paris, 1982.
28. M. Parikh and D. Kyser, *J. Appl. Phys.* **50**, 1104 (1979).
29. N. Eib, *J. Vac. Sci. Technol. B* **1**, 425 (1985).
30. S. Rishton and D. Kern, *J. Vac. Sci. Technol., B* **1**, 135 (1987).

Chapter 5

Ion-Beam Lithography

BENJAMIN M. SIEGEL

School of Applied and Engineering Physics
Cornell University
Ithaca, New York 14853

I. INTRODUCTION

The endeavor to structure to ever smaller dimensions has generated growing interest and investigations in ion-beam lithography (IBL). This chapter evaluates the potential of using ions in the microfabrication process, especially in the submicrometer and nanometer range of dimensions. We shall discuss the physics of the various processes and their limitations to obtain an understanding of the present state of the art. Ion-beam lithography cannot be considered a technology at this stage of its development, but with a very large amount of research and development in the field it can be expected to contribute significantly to the structuring and fabrication of very high resolution experimental devices, an area of investigation that is of fundamental interest. This subject has been discussed by Howard and Prober in Chapter 4, Volume 5 of this treatise, *VLSI Electronics: Microstructure Science* [1].

The lithographic process, a method of defining patterns in microfabrication, involves writing the desired pattern, generally in a resist, developing this pattern, and transferring it in some form to the structure or device to be fabricated. While the current technology uses primarily light and UV radiation to expose the resist in this writing process, focused electron beams and x-rays are used in some special applications in the technology of microfabrication. Ion beams are still in an early stage of development. A focused probe of light ions such as H^+ would provide the best ion beam for direct writing in resists [2] in a manner analogous to electron-beam lithography (EBL), as we shall see when ion–matter interactions are discussed. However, beams of heavy ions generated with liquid metal ion sources [3,4] are under extensive development for several applications, for example, direct implantation of different ion species for doping of microelectronic devices [5a,b,6], mask repair [7], and ion-beam-enhanced reactive ion etching [8]. Therefore, in this chapter we also discuss liquid metal ion sources and how they are applied to microfabrication processes.

Ion-beam exposure of resists in the lithographic process has the possibility of very high resolution structuring because the ions deposit their energy in a very narrow range, producing secondary electrons of very low energy and therefore short range. Thus, there is no backscattering and minimal proximity effects, so high-density architecture can be patterned in the devices for VLSI. This possibility is especially true for exposure with the lightest ion such as H^+. The nature of the energy deposition from ions of different mass and energy in the resist determines the characteristics of the patterns that can be written and transferred in the lithographic process. The consequences of the fact that an ion deposits some 100 times more energy than an electron in a smaller volume [9a,b] is the basis for much of the interest in the development of IBL as the method needed for structuring at nanometer dimensions.

While broad beams of ions have been in general used for ion implantation, etching, milling, plating, etc., focused ion beams have not been available until recently for the lithographic processes. It was only with the development of high-brightness sources that it became practical to use ion beams for direct writing in resists or structuring in substrates. The development of the liquid metal ion (LMI) source [3–5b,10–12] provided the first basis for obtaining a high-resolution, high-current-density focused ion probe that could be used to write patterns in resists in the same manner that electron beams are used in the lithographic process. More recently, gaseous field ionization sources have been developed, an H_2^+ source in particular [13] that has a very high brightness and provides the basis for very high resolution structuring in resists with a light ion.

In this chapter particular attention is paid to the two factors that underlie the advantages of IBL and the necessary conditions for achieving high-resolution structuring with IBL: the nature of the energy deposition in resists or substrates as functions of ion mass and energy and the characteristics of the ion sources that make high-resolution IBL feasible.

II. PROCESSING WITH ION BEAMS

A. The Lithography Process

Lithographic processes are used to define patterns in the semiconductor materials employed in the microfabrication of VLSI devices. There are a number of reviews of patterning and structuring by lithography as currently applied in microfabrication [14–18]. Readers not familiar with the basic processes are referred to these discussions, particularly those concerning EBL, in which many of the same considerations as those in IBL are involved. We shall be interested in the processes that produce very high resolution patterns and the factors that set the limits to the resolution in the range of nanometer structures [1], for it is in these dimensions that IBL can provide new possibilities.

The lithographic process generally involves patterning a layer of radiation-sensitive material, referred to as a resist, that has been deposited on a substrate, for example, poly(methyl methacrylate) (PMMA) on a silicon wafer. The resist layer is exposed in the desired pattern to radiation — photons, electrons, or ions — which changes the chemical or physical characteristics of the layer. The radiation either makes the resist more (positive) or less (negative) sensitive to the development processes by which the exposure pattern is transferred to the substrate material. The development processes can involve either liquid solution or dry plasma etching of the resist. Of course, the areas of the polymer that are not removed in the development step must be resistant to the solvent or etching procedure — hence, the term *resist* for the layer used for the stencil pattern. We shall be interested in the methods of producing very high resolution stencils that provide the basis for subsequent pattern transfer to the substrate. The pattern transfer involves either the deposition of material by evaporation or ion implantation on the substrate through the open areas of the resist or the structuring of the substrate by an etching process. The etching processes usually involve some form of dry etching such as reactive ion etching [18]. Of course, with focused ion beams of different ion species, doping can be done by direct implantation without a stencil pattern in a resist layer [5a,b,6]. The same holds for etching patterns directly in the substrate [19a,b], although the etching rate is greatly enhanced if a reactive gas such as chlorine is present [20] (see Fig. 1 for schematic representations of these processes). The methods considered in this chapter are related to very high resolution IBL, its potential and limitations, and these considerations will be compared with the characteristics that are possible with EBL.

B. Pattern Generation by Ion-Beam Lithography

Ion-beam exposures of resists can be made by either parallel or serial methods using broad ion beams or ion beams focused to high-resolution probes, respectively. In parallel processing the resist is exposed through a mask with a projection system or by proximity printing. While these methods have the potential for very fast throughput and could compete with x-ray projection printing, there are limitations on the resolution and flexibility that can be achieved. Serial exposure methods, which use a focused probe of ions scanned vectorially or in a raster, offer the possibility of critical alignment, correction for wafer distortion, and easy modification of the exposure pattern features. All of these capabilities are highly desirable at this experimental and developmental stage in IBL. Focused-ion-beam writing has the potential of producing very high resolution patterning with high-density structures and complex architectures, because line widths and spacings are minimally limited by proximity effects. At this stage these capabilities are much more important than very high throughput rates. Given the very high brightness ion sources now available, the patterning rates with focused ion probes may ultimately be comparable to or exceed the rates available with focused electron beams.

Focused ion beams can be used to produce patterns in some unique

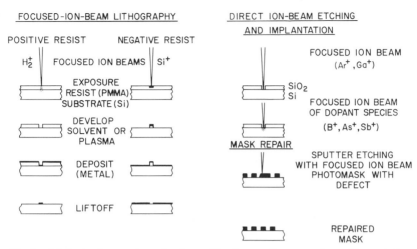

Fig. 1. Methods of patterning using focused ion beams. Exposure of a positive resist (e.g., PMMA) with development, metal deposition, and liftoff. Implantation of Si$^+$ ions in a resist (e.g., PMMA) to produce a negative resist effect. Direct implantation of dopant through a stencil mask produced by focused-ion-beam sputter etching. Mask repair by sputter etching out unwanted metal areas. Masks could also be repaired by ion deposition to fill holes in masks.

ways. Ions such as Si$^+$ can be directly implanted in a resist such as PMMA to produce areas that are much more resistant to ion plasma etching [21], or the focused ion beams can be used to write patterns in substrates by ion-beam-assisted etching [22a,b].

III. ION–SUBSTRATE INTERACTIONS

The interactions of intermediate energetic ions (20–200 keV) with the materials used in the lithographic process for the fabrication of microelectronic devices have characteristics that are significantly different from those of the other radiations used to expose resists. Ions have larger effective masses with greater cross sections of interaction with the atoms in a solid, so they deposit their energy in a much smaller volume and penetrate the resist or substrate (or both) in a more limited and well-defined range. These characteristics provide the potential for much higher resolution and greater sensitivity in structuring by IBL processes than in any other method except, perhaps, x-ray lithography. However, x-ray lithography requires the use of masks, and the very high resolution masks needed could be made by direct write IBL. A review of the basic mechanisms involved in ion–

substrate interactions will help the reader to understand and appreciate the ways in which ion beams can be used in high-resolution structuring and microfabrication.

A. Primary Interactions

A complex variety of ion–substrate interactions and secondary effects are involved when any beam of energetic ions is incident on a resist layer or other solid surface. The major primary and secondary interactions of ions of intermediate energy (20–200 keV) incident on solids are listed in Table I. Of the several processes that occur, the one of greatest interest for understanding the exposure of resist by ion beams is (1) the energy transfer process that causes chemical bonds to break or cross-link and alters the characteristics of the exposed resist as required for the subsequent development processes. Among the other interactions is the transfer of the momentum of the incident ion to the surface atoms in the solid, dislodging them from weakly bound positions so either these atoms are (2) relocated or, in the case of incident ions with greater energies, enough momentum is transferred to cause the atoms to be freed, (3) producing a sputtering effect that could be an important method for direct microstructuring with focused ion beams. Or the ions can penetrate into the substrate lattice, (4) causing atomic dislocations and this radiation damage of the lattice can be utilized in subsequent processing steps in the microfabrication process (e.g., reactive ion etching). The incident ions can also become trapped in the lattice, (5) effectively implanting given ion species in the substrate to

TABLE I

Ion–Solid Interactions: Intermediate Energy (20–200 keV) of Incident Ions

Primary transfer in solid	Secondary effects
Ion stopping by electrons of atoms in solid	Low-energy secondary electrons
Break chemical bonds	Break chemical bonds
Ion nuclear stopping	Photon emission
Internal dislocations	X-ray emission
Break chemical bonds	Auger electrons
Ions implanted in solid	
Ion–atom scattering	
Physical sputtering	
Surface dislocation	
Charge transfer	
Ionized surface atom emission	
Backscattering	

produce the desired doping of the electronic materials or to change the physical or chemical characteristics of the implanted material in a desired manner.

B. Secondary Effects

A variety of secondary processes follow the interaction of intermediate-energy ions with a solid. One is (6) secondary electron emission, a particularly useful phenomenon since these secondary electrons can be utilized to observe the surface structure being scanned by the focused primary ion beam. This process is analogous to that used in the scanning electron microscope in which one of the detected signals used for imaging is also the secondary electron emission produced by a focused electron probe incident on the surface whose topography is to be observed. Characteristic x-ray emission also occurs when inner shell electrons in the substrate material are ionized. The decay process usually involves (7) the emission of the characteristic x-ray along with (8) the production of Auger electrons. These secondary emissions of x-ray photons and Auger electrons could be used for characterizing the atomic species of the substrate and provide the possibility of microanalysis with very high resolution. Another secondary excitation process involved in the energy transfer of the ion energy to the solid may (9) produce photon emission by energy decay through a number of atomic and lattice mechanisms. A very important and widely applied effect related to (2) above is the secondary emission of the ions of the various atomic species in the substrate. These ions can be mass analyzed and provide a powerful microanalytical tool known as scanning ion mass spectroscopy (SIMS) for studying the atomic composition of the near surface of solids.

C. Energy Transfer Processes

The energy transfer processes that play primary roles in stopping the ions as they penetrate the solid are the interaction of the ions with the electrons of the lattice atoms and the interaction of the ions with the nuclei of the atoms in the solid. The characteristics of these interactions determine the distribution of the energy deposited in the resist or substrate (or both) and therefore the subsequent profiling in the resist or substrate in the lithographic process as well as the processes involving ion implantation, radiation damage, etc. The characteristics of the electron stopping and nuclear stopping processes can be treated separately, and the amount of each type of energy loss as the ion penetrates the substrate depends primarily on the velocities of the incident ions and the relative masses of the incident ions to the substrate atoms.

The stopping of ions in solids, both theoretical and experimental, has attracted the attention of many investigators. Ziegler [23a] has written a good short review with extensive references of both the early work on this subject and the current status in this field. The basis of the theory used today and known as the LSS theory was presented in a paper by Lindhard *et al.* [24], with later papers detailing some of the equations of the theory [25a–c]. These papers developed the earlier approaches and developed them into a unified theory of stopping power and range based on statistical models of atom–atom collision. Improvements in this theory have been made since the mid-1960s through the development of some better approximations as well as numerical methods that take advantage of the power of the computers now available. Stopping power and range distribution in amorphous targets can now be calculated with an average accuracy of 2 to 10% depending on the energy and atomic number of the incident ions [23a].

1. Electron Stopping Cross Sections

The LSS theory is based on the local-density approximation and treats the interaction of the incident ion as a perturbation on the electron gas of the solid. The electronic stopping S_e can be given as

$$S_e = \int I(v_1 \rho)(Z_1(v))^2 \rho \, dV, \tag{1}$$

where I is a stopping interaction function of a particle of unit charge with a velocity v interacting with a volume dV of the target that has a free electron gas of density ρ, and $Z_1(v)$ is the effective charge on the incident ion, where the charge may differ from the atomic number because the ion may not be fully stripped. The theory assumes the solid to have a slowly varying electron plasma density, so the charge of the particle interacts with independent single volume elements of plasma. By taking these interactions over all the volume elements and given the mean interaction of the ion with the solid, the final stopping power is obtained.

a. Light Ions Incident on Solids. The interaction of a charged particle (e.g., H^+) with ion energies of 100 and 10,000 keV/amu incident on a target of copper is illustrated in Fig. 2 [23a]. The dotted line is the charge distribution of electrons in the solid copper based on a Hartree–Fock model. The dashed lines show the interaction of the charged particle with that electron distribution. The solid line shows the stopping power S_e [Eq. (1)] integrated over the volume of the solid. In the case of the 100-keV ion

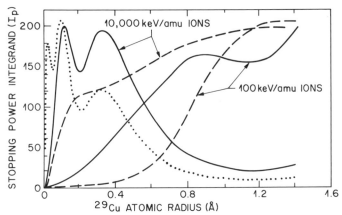

Fig. 2. Interaction of 100- and 10,000-keV particles with copper. The dotted line is the charge distribution in ^{29}Cu given by the Hartree–Fock model of atoms in a solid. The dashed curves are the interaction I of the respective particles with this electronic charge. The solid lines are the integrated stopping powers S_e [Eq. (1)] showing the electronic energy transfer to the copper as charged particles with low and high energy are stopped in the solid. The lower-energy ions interact appreciably only with the conduction electrons. These curves are a composite of curves taken from J. F. Ziegler, *in* "Ion Implantation Science and Technology" (J. F. Ziegler, ed.), pp. 83–85. Academic Press, London, 1984.

there is almost no interaction of this relatively slow ion with the inner K-shell electrons and very little with any electrons except with the conduction electrons. For interaction to take place the velocity of the incident ion must be greater than the electron velocity in the solid. When the ion velocity is at the Fermi velocity or below that of the electrons in the solid, the stopping interaction I falls off. Ninety percent of the energy of the ion is absorbed by the interactions with the outer shell electrons. Very few x rays are produced by this interaction because there is so little excitation of K- and L-shell electrons. The characteristics of the interaction are different for higher-energy protons; for example, 10 MeV/amu protons, with their higher velocities, do interact with the lower shell electrons, particularly the L-shell electrons, and produce x-ray emission, as shown by the solid line for that energy in Fig. 2.

Extensive calculations on the electronic stopping cross sections for protons were made by Ziegler and collaborators [26]. To obtain the electronic cross section for He$^+$ ions the equivalent H$^+$ stopping is multiplied by the effective charge of the helium at that velocity, $S_{He} = S_H(Z_{He}\gamma_{He})^2$, where γ is the effective fractional charge [27].

b. Heavy Ions ($Z > 2$) Incident on Solids. The electronic stopping of heavy ions in solids is strongly dependent on the relative velocities of the ion v_1 and the Fermi velocity of the electrons of the atoms in the target v_F. It is convenient to divide the interaction into three velocity ranges: (1) low-velocity ions ($v_1 < v_F$), or <25 keV/amu); (2) an intermediate velocity ($v_F < v_1 < 3v_F$, or 25–200 keV/amu); and (3) a high-velocity regime ($v_1 > 3v_F$, or >200 keV/amu). Since the Fermi velocity is nearly equal to the Bohr velocity v_0, we can represent each of these ranges in kiloelectron volts per atomic mass unit.

In the low-velocity range Linhard and Scharff assumed an ion interacting with a uniform electron gas [24,28] and found that the energy is transferred proportionally to the relative drift velocity of the electron with respect to the ions. Later, more accurate calculations still yielded velocity-dependent electronic stopping cross sections. An important exception is for the case of ions incident on semiconductor targets, which do not have a free electron gas as do metal targets. While slow ion–metal target interaction gives a satisfactory approximation with theory, a correction based on empirical data must be used to obtain satisfactory correlation with theory for semiconductor targets. Experimental data on the stopping power of low-velocity ions in silicon and germanium give a relation S_e proportional to $v^{0.7}$ for incident ions with atomic number less than 19 [23a]. Since the band gap in semiconductors allows fewer low-energy excitation levels, a lower energy loss is to be expected.

In the high-velocity range the stopping power S_{HI} for a heavy ion can be scaled to the equivalent proton stopping power just as helium stopping was scaled,

$$S_{HI} = S_H Z_{HI}^2 \gamma^2, \qquad (2)$$

where Z_{HI} is the atomic number of the heavy ion and γ its fractional effective charge. The effective charge can be obtained to a good approximation from the Thomas–Fermi atomic theory for a wide range of heavy-ion stopping.

In the intermediate range the treatment of Brandt and Kitagawa [29] can be used. They modified previous treatments of the theory by taking the effective charge state of the ion as based on its charge state in a solid, assuming the stripped ion has only electrons whose velocities are greater than the relative velocities of the ion to the Fermi velocity of the solid. In this velocity range it is also necessary to treat close collisions separately. Here the electrons of the target atoms penetrate the electron shells of the ion, so there is less shielding of the nucleus and an increased energy loss occurs.

2. Nuclear Stopping Cross Sections

Nuclear stopping depends primarily on the energy of the incident ion and the ratio of the mass of the ion to the mass of the target atoms. There is a wide range in the amount of nuclear stopping between light and heavy ions. Nuclear interactions produce strong angular scattering and straggling, so the energy deposition in the solid is in a much larger width and volume, giving poorer profiles for high-resolution lithographic processing. Only a very small fraction ($\frac{1}{100}$) of the energy lost by 100-keV H^+ ions is through nuclear interactions, so H^+ stopping in this energy range can be treated as electron stopping. On the other hand, for the heavy ion Ga^+ (100 keV), for example, 80% of its energy is lost through nuclear stopping.

The nuclear stopping of energetic ions incident on solids involves elastic collision scattering of two atoms with energy transferred to the stationary atoms of the solid. To calculate the kinematics of this interaction the atom–atom interatomic potentials are reduced to a single universal analytical function. This function is used to generate new universal stopping cross sections and scattering functions for calculating any combination of ion–solid penetration. The interaction involves the Coulombic interaction of the positive point charges of the nuclei screened by their electrons, which reduce their values at different radii. This subject has been reviewed by Ziegler [23a], who has also presented a more recent, fuller treatment with collaborators [30a]. The interatomic screening function is defined as

$$\Phi_1 \equiv \frac{V(r)}{Z_1 Z_2 e^2 / r}, \tag{3}$$

where $V(r)$ is the atomic potential as a function of radius r; Z_1 and Z_2 are the atomic numbers of the incident ions and substrate atoms, respectively; and e is the electronic charge. Ziegler et al. [30a] have obtained a reduced radial coordinate,

$$a_u = 0.8854 a_0 / (Z_1^{0.23} + Z_2^{0.23}), \tag{4}$$

which allows the development of an analytical function that can be used to calculate accurately the interatomic potential,

$$\Phi_u = 0.1818 e^{-3.2x} + 0.5099 e^{-0.9423x} + 0.2802 e^{-0.4028x}$$

$$+ 0.02817 e^{-0.2016x}, \tag{5}$$

where the reduced radial separation is $x = r/a_u$. This universal screening function is plotted in Fig. 3.

To obtain the energy transferred from the incident ion to the substrate per unit path length, dE/dR, the nuclear stopping cross section $S_n(E)$ must be obtained, since

$$dE/dR = NS_n(E), \tag{6}$$

where N is the atomic density of the target. The reduced universal nuclear stopping S_n as a function of reduced energy E given by Ziegler [23a] is also

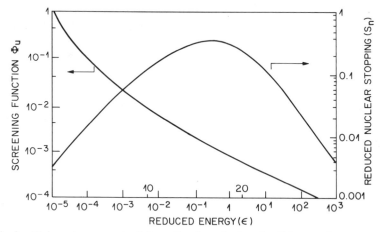

Fig. 3. Universal screen potential as a function of reduced radial separation $x = r/a_u$ and the universal nuclear stopping power as a function of reduced energy [Eqs. (7)–(10)]. This figure is a composite of curves taken from J. F. Ziegler, *in* "Ion Implantation Science and Technology" (J. F. Ziegler, ed.), pp. 71–73. Academic Press, London, 1984.

plotted in Fig. 3. For calculations in practical units,

$$S_n(\epsilon_0) = \frac{8.462 \times 10^{-15} Z_1 Z_2 M_1 S_n(\epsilon)}{(M_1 + M_2)(Z_1^{0.23} + Z_2^{0.23})} \quad \frac{eV}{atom/cm^2}, \tag{7}$$

where the reduced energy is given by

$$\epsilon = \frac{32.53 M_2 E_0}{Z_1 Z_2 (M_1 + M_2)(Z_1^{0.23} + Z_2^{0.23})}, \tag{8}$$

and the reduced nuclear stopping is given by

$$S_n(\epsilon) = \frac{\ln(1 + 1.1383\epsilon)}{2(\epsilon + 0.01321\epsilon^{0.21226} + 0.19593\epsilon^{0.5})} \quad \text{for } \epsilon \le 30 \tag{9}$$

and

$$S_n(\epsilon) = \ln(\epsilon)/2\epsilon \quad \text{for } \epsilon > 30, \tag{10}$$

where at the higher energies the scattering becomes like Rutherford scattering and the universal nuclear stopping more accurately approximates the physical effects.

3. Ion Range and Energy Distribution in Solids

In IBL and ion implantation, the range of the incident ions in the resist or substrate (or both) and the distribution of their energy as they are stopped are the parameters of primary interest. The analytical methods based on the LSS theory that have been discussed above can be used to obtain the data. However, computer simulation using a Monte Carlo type of calculation has become the more powerful, preferred method for obtaining these parameters. Monte Carlo methods make use of the same basic ion–substrate interactions but can consider more complex mechanisms, within the limits set by the computer, to obtain the characteristics of the penetration, energy deposition, and other desired parameters. Monte Carlo methods have the advantage of following each incident ion through the substrate and can determine penetration through interfaces of different materials. Of course, to obtain statistically significant results the trajectories of a large number of ions must be calculated and that can require considerable computer time. However, Monte Carlo techniques have been used in a number of applications that require the simulation of the scattering processes in solids, for example, ion implantation, sputtering, radiation damage, as well as writing in resist material for IBL.

a. Light Ions Incident on PMMA Resist. A Monte Carlo computer program for determining ion range, energy, and damage distribution called TRIM was developed by Biersack and Haggmark [30b] and is described in detail and used to obtain the data given by Ziegler *et al.* in their monograph [30a]. Another Monte Carlo program was developed by Adesida and Karapiperis [31] and Karapiperis [32a]. Named PIBER (program for ion beam exposure of resist), it was used to investigate the projected range R_p, mean path length R, standard deviation of the range distribution σ, ion reflection coefficient R_N, and depth and width of the energy deposited in the target material. These investigators were particularly concerned with ions incident on PMMA resists to determine the applicability of focused ion beams to IBL.

Figure 4 shows the trajectories obtained by Karapiperis *et al.* [32b] with Monte Carlo calculations of 60-keV H^+ ions incident on a 4000-Å PMMA layer on a silicon substrate and on a 4000-Å PMMA layer on gold. The characteristics of the interactions, the penetration, and the energy deposition are very clear. The proton interaction with the PMMA is essentially electronic and there is very little lateral scatter, so the ion energy is deposited in a narrow band. Only near the end of the ion trajectories in the silicon or gold layers when ions are moving slowly are there appreciable ion–nuclear interactions with large-angle scattering or backscattering.

Adesida *et al.* have carried out Monte Carlo calculations and experimental determination of several light ions (H^+, He^+, Li^+, Be^+, B^+, C^+) in PMMA with energies ranging from 5 to 300 keV [33,34]. PMMA was bombarded with the different ions and developed in a solvent that dissolved the exposed resist much faster than the unexposed areas. The satu-

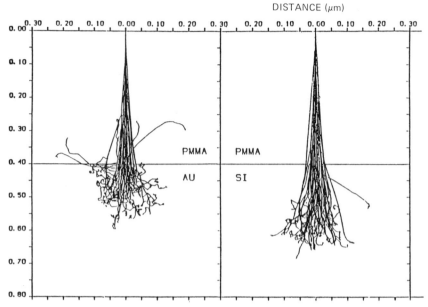

Fig. 4. Trajectories of 60-keV H^+ ions in PMMA on substrates of gold and silicon calculated using the PIBER program. [From L. Karapiperis, I. Adesida, C. A. Lee, and E. D. Wolf, *J. Vac. Sci. Technol.* **19**, 1260 (1981).]

rated developed depth was taken to be the mean path length of the incident ion. They fitted the experimental data to the Monte Carlo calculations to obtain the projected range and straggles of the different ion species. In fitting the data it was necessary to introduce a corrected value of the LSS electronic stopping power.

Adesida *et al.* used $S_L = K_L E^{1/2}$, where

$$K_L = Z_1^{1/6} \left(\frac{0.0793 Z_1^{1/2} Z_2^{1/2} (A_1 + A_2)^{3/2}}{(Z_1^{2/3} + Z_2^{2/3})^{3/4} A_1^{3/4} A_2^{1/2}} \right)$$

Here, Z_1 and Z_2 are atomic numbers of the incident ion and target, respectively, A_1 and A_2 are their respective atomic weights, and E is the energy of the incident ion. The correction was made by using a scaled value of K, $CK = K/K_L$. Other investigators have used the saturated development depth in resist to measure the mean path length of some ion species in PMMA, this being the resist of choice because its development characteristics are well known. Ryssel *et al.* [35] used this technique with H^+, He^+, Ar^+, and Ga^+ ions. Moriwaki [36] measured only H^+ in PMMA. Plotted in Fig. 5a are the results of Adesida and Karapiperis [34] for H^+ incident on PMMA, which give the mean path length R, projected range

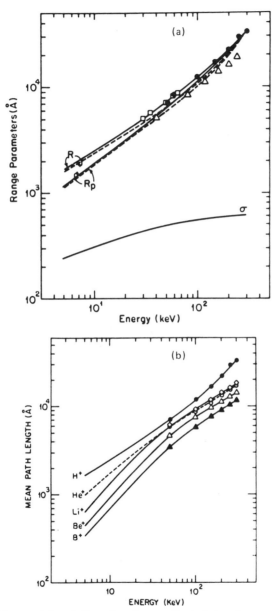

Fig. 5. (a) Mean path length R, projected range R_p, and straggle σ versus energy of H^+ ions implanted in PMMA resist (——, PIBER (CK); ---, PIBER (AZ); ●, experimental, Siegel; □, experimental, Meriwaki; △, experimental (Ryssel *et al.*). (b) Mean path length for various ions implanted in PMMA. [From I. Adesida and L. Karapiperis, *J. Appl. Phys.* **56**, 1805 (1984).]

R_p, and straggle as a function of the incident energy of the ions. The solid line represents their data with the scaled correction factor CK. Their results are also compared with the data they obtained using the PIBER program but with the electronic stopping power given by Andersen and Ziegler (AZ) [26] (dashed line). The experimental data of Ryssel *et al.* [35] and Moriwaki [36] are also plotted in this figure, showing the best agreement with the data of Moriwaki. Figure 5b is a composite plot of the calculated and experimental data Adesida *et al.* [33,34] obtained for the mean range of several light ions incident on PMMA at energies from 5 to 300 keV.

b. Heavy Ions Incident on PMMA. The trajectories of 250-keV Ga$^+$ ions in a 2500-Å layer of PMMA on silicon were obtained with Monte Carlo calculations by Karapiperis [32a]. These trajectories are illustrated in Fig. 6. The penetration depth of the 250-keV Ga$^+$ ions is considerably lower than that of the 60-keV H$^+$ ions even though they have a much higher energy. The stopping of the ions is primarily nuclear with these heavy ions, so there is wide-angle scattering as well as longitudinal straggling. The lateral spreading of the beam deposits energy in the resist in a

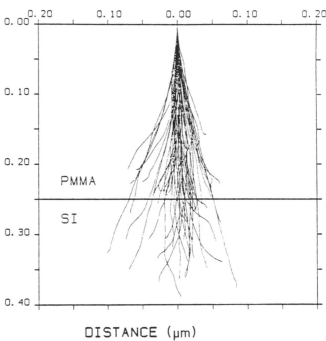

Fig. 6. Trajectories of fifty 250-keV Ga$^+$ ions in 2500-Å-thick PMMA on silicon. [From L. Karapiperis, Ph.D. Thesis, p. 119. Cornell Univ., Ithaca, New York, 1982.]

much wider region than the incident beam, while the longitudinal straggle produces a range of penetration that is dose dependent.

   c. Lateral Spreading of Light and Heavy Ions in PMMA. Of importance for the determination of the resolution limits that can be realized by direct writing with focused ion beams in resists is the lateral spreading of the ion trajectories in the resist layers and in the substrates on which the resists are deposited. Figure 4 shows the trajectories of 60-keV H^+ ions penetrating PMMA on gold and silicon substrates. It is evident from this figure that there is a very small spread of the ions at depths of 4000 Å, and

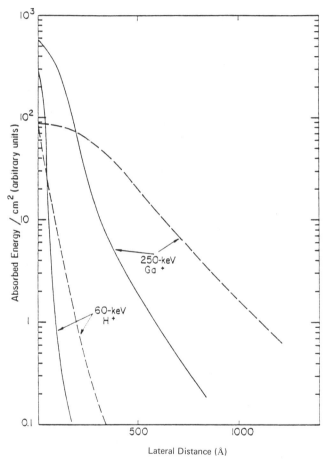

Lateral Distance (Å)

Fig. 7. Absorbed energy versus lateral distance away from a δ-line exposure at two different depths for 60-keV H^+ and 250-keV Ga^+ into PMMA (——, depth = 1200 Å; ---, 2800 Å). [From L. Karapiperis, Ph.D. Thesis, p. 120. Cornell Univ., Ithaca, New York, 1982.]

backscattering is absent from the silicon substrate and very small from the gold substrate. For comparison the trajectories of 250-keV Ga^+ ions in PMMA on silicon are shown in Fig. 6 [32a]. The lateral spreading is much greater, and therefore the energy deposited in the resist extends over a much wider distance from the delta function of incident ions. The energy deposited per square centimeter for 60-keV H^+ ions and 250-keV Ga^+ ions is shown for two depths in the resist layer in Fig. 7 [32a]. We shall see the important consequences of these data and why the lightest atoms are best for patterning in resists with focused ion beams.

D. Modeling of Exposure and Development of Ions Incident on Resists

1. Hydrogen Ions in PMMA

Karapiperis [32a] and Karapiperis *et al.* [32b] have also carried out modeling experiments based on their Monte Carlo calculations as inputs to a convolution and development program developed by Neureuther and Rosenfield [37a,b]. The absorbed energy distributions obtained from the Monte Carlo calculations (PIBER) are convoluted with realistic ion-beam shapes such as Gaussian distributions, square profiles, or aberrated image figures. A string development model is used to simulate developed profiles in the resist. This program of *con*volution and *dev*elopment (CONDEV) modeling has been used by Karapiperis [32a] and Karapiperis *et al.* [32b] to obtain the contours that would be obtained in a 4000-Å layer of PMMA exposed to a rectangular beam of 60-keV H^+ ions 1000 Å wide with dosages of 0.8 and 1.5 1 $\times$ 10^{-6} C/cm². One-half of the developed contours are illustrated in Fig. 8 after different development times in 1 : 1 MIBLK – IPA. Breakthrough occurs after 2.5 min, but since the energy deposition is confined to a narrow band even at the full depth of the resist the contour changes very little after 7.5 min, and even after 17.5 min has broadened by only 250 Å. To demonstrate the minimal proximity effects to be expected on exposure of PMMA by a focused H^+ ion beam, these investigators have used the CONDEV program to model the contours obtained when 4000-Å-thick layers of PMMA are exposed to a pattern composed of five lines of 60-keV H^+ ions each having profiles 1000 Å wide and spaced 1000 Å apart. The modeled profiles obtained for an exposure dose of 1.5 $\times$ 10^{-6} C/cm² at various development times are shown in Fig. 9.

While the effects of secondary electrons were not considered in the PIBER program, the range of the secondary electrons should be limited. The maximum energy that can be transferred by 60-keV H^+ ions is some 100 eV, and the maximum range of any such secondary electrons is

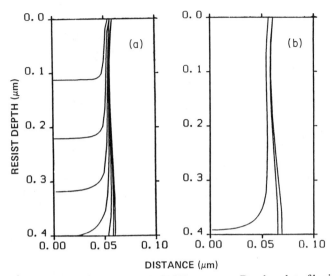

Fig. 8. Modeling using PIBER and CONDEV programs. Developed profiles in PMMA. Incident rectangular cross-section beam of 60-keV H^+ ions. Beamwidth 1000 Å. Doses: (a) 0.8×10^{-6} C/cm^2; (b) 1.5×10^{-6} C/cm^2. Development times: (a) 1, 2, 3, 4, 5, 6, 7 min; (b) 1, 4, 7 min. [From L. Karapiperis, Ph.D. Thesis, Cornell Univ., Ithaca, New York, 1982.]

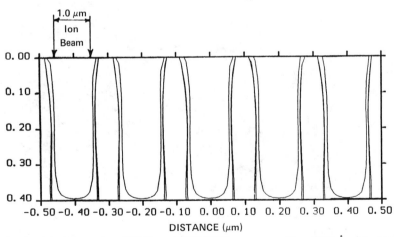

Fig. 9. Modeling using PIBER and CONDEV programs. Five 1000-Å-wide parallel lenses, 1000 Å apart, developed in PMMA. 60-keV H^+ ions; dose, 1.5×10^{-6} C/cm; developed in 1, 4, and 7 min. [From L. Karapiperis, Ph.D. Thesis, Cornell Univ., Ithaca, New York, 1982.]

~ 50 Å, mostly in the forward direction, so the lateral spreading should be confined to a few tens of angstroms. These factors explain the narrow profiles that do not broaden more than 250 Å at the deepest level even with considerable overexposure and overdevelopment. Therefore, the proximity effects are negligible at this level of resolution, and high-contrast profiles are obtained over a wide latitude of exposure and development.

High-resolution focused probes of H$^+$ ions have not yet been demonstrated, but in recent experimental investigations PMMA was exposed through stencil masks and the high-resolution potential of IBL was demonstrated. Kumoro *et al.* [9a], Karapiperis and Lee [38], Zhang *et al.* [39], and Adesida *et al.* [40] used masks to expose PMMA resists to very narrow lines of H$^+$ ions. Figure 10 shows the lines obtained by Adesida *et al.* [40] and the stencil mask of gold–palladium-coated silicon nitride fabricated with the aid of a scanning transmission electron microscope (STEM).

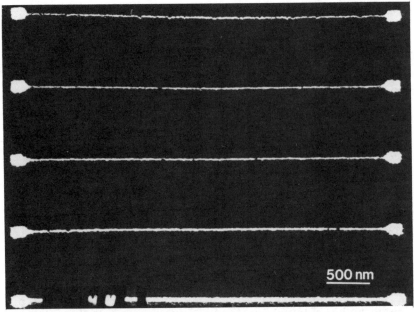

Fig. 10. (a) Bright-field STEM micrograph of a stencil mask. Feature size ranges from 16 to 100 nm in fabricated masks. (b) Bright-field STEM micrograph, 80-nm line. (c) Bright-field STEM micrograph, 30-nm line. These lines were replicated by exposing 70-nm-thick PMMA with 30-keV H$^+$ ions through the stencil mask illustrated in (a). The lines are etched 30 nm deep into the 50-nm Si$_3$N$_4$ substrate. (Dark spots in background are gold particles used for focusing.) [From I. Adesida, E. Kratschmer, E. D. Wolf, A. Muray, and M. Isaacson, *J. Vac. Sci. Technol.*, B **3**, 45 (1985).]

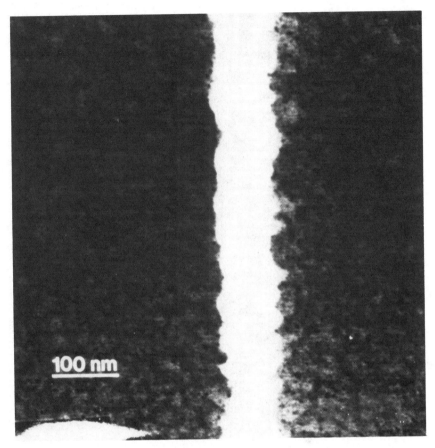

100 nm

Fig. 10 *(Continues)*

2. Gallium Ions in PMMA

The same conditions do not obtain if the incident ions are heavy ions of high atomic number. Heavier incident ions are stopped by nuclear interactions as they penetrate the substrate, and this process involves considerable wide-angle scattering of the ions as they traverse the target. Karapiperis's [32a] Monte Carlo calculations on 250-keV Ga^+ ions incident on a 2500-Å-thick resist layer of PMMA on silicon are illustrated in Fig. 6. Nuclear stopping accounts for some 85% of the energy loss in this case, so there is substantial lateral straggling as well as longitudinal straggling. These factors introduce important limitations on the use of heavy-ion beams such as Ga^+

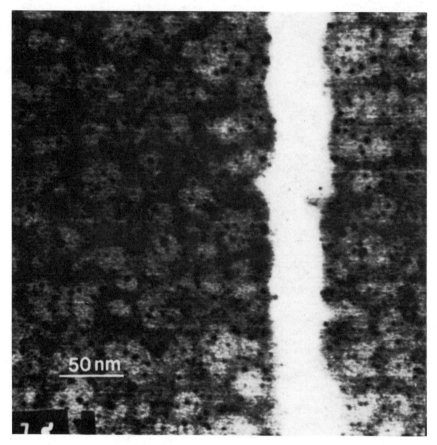

Fig. 10   *(Continued)*

for exposure of resists in the lithographic process by direct writing that can be deduced from the calculation of the absorbed energy per square centimeter versus lateral distance at two depths for both 60-keV H^+ and 250-keV Ga^+ ions (Fig. 7). The much wider lateral spread of the energy deposition in the case of the Ga^+ ions is evident. The range of the Ga^+ ions is only about half that of the H^+ ions, even with the higher energy to which the Ga^+ ions are accelerated. The longitudinal straggling of the Ga^+ ions also produces a range of penetration that is dose dependent and results in saturation development depths that are dose dependent. Karapiperis [32a] also used the CONDEV program to model the results of exposing a 0.3-μm-thick layer of PMMA to a 0.1-μm-wide Ga^+ ion beam at 250 keV energy with a dose of 0.1×10^{-6} C/cm^2. The resulting contours after

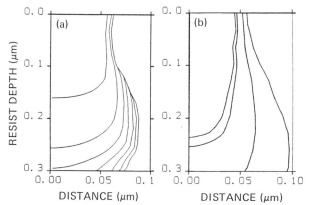

Fig. 11. Modeling by PIBER and CONDEV. Developed contours in PMMA exposed by a 1000-Å line of 250-keV Ga$^+$ ions. (a) Dose 0.10×10^{-6} C/cm^2; development times $t = m + 2.33$ min ($m = 0, 1, \ldots, 7$). (b) Equi-energy contours at 800, 600, 200, and 40 J/cm^3. [From L. Karapiperis, Ph.D. Thesis, p. 137. Cornell Univ., Ithaca, New York, 1982.]

different development times are illustrated in Fig. 11. The broadening of the line with development times follows the spread of absorbed energy. Figure 12 shows the results obtained with the CONDEV program using an input of five 1000-Å-wide lines of 250-keV Ga$^+$ ions spaced 1000 Å apart. The wide broadening of the lines with development time produces the equivalence of a "proximity" effect and would surely limit the application of such a heavy-ion beam for very high resolution direct writing. Heavy ions of lower energy could be used to expose thin top layers of multilayer resists, but this procedure would lose much of the potential advantages available with IBL.

As we have seen, in exposures of PMMA to H$^+$ ions in which the energy

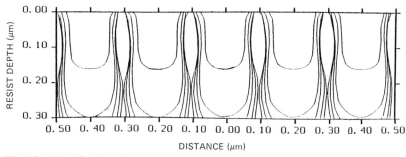

Fig. 12. Modeling by PIBER and CONDEV. Development profiles in PMMA produced by five 1000-Å-wide lines of 250-keV Ga$^+$ ions spaced 1000 Å apart at development times of $t = 2m + 2.33$ min ($m = 0, 1, \ldots, 4$). [From L. Karapiperis, Ph.D. Thesis, p. 138. Cornell Univ., Ithaca, New York, 1982.]

loss is by electronic stopping, the characteristics of the process are much more satisfactory for exposures of resists in IBL. The penetration range is dependent on the incident energy of the H^+ ions, and saturation development depth is dose independent. The very minimal undercutting would allow very high density patterns to be exposed in the resists.

E. Ion-Beam Resists

The nature of the ion–resist interactions provides possibilities of new approaches to resist characteristics and types of resists that can be used in IBL. Typical organic resists are some two orders of magnitude more sensitive to ions than to electrons. As we have seen, the energy of the incident ion, at least in the case of light ions, is deposited in a very narrow range, both laterally and longitudinally. Also, there is no appreciable backscattering from the substrate as is the case in electron exposure of resist. Secondary electrons have low energy ($\simeq 50-100$ eV) so they will not broaden the energy deposited by the primary-ion exposure. Nor will the secondary electrons that may be backscattered from the substrate interface, for their energy is very low ($\simeq 15$ eV). These characteristics provide the possibility of tailoring resists to optimize the resist material for processing considerations other than sensitivity, for example, adherence to the substrate and integrity under the subsequent processing involved in pattern transfer. The potential for obtaining high-resolution, high-contrast resists with superior processing qualities is very promising. However, the minimum continuous linewidth that can be achieved will still be set by resist contrast, the energy profile of the beam, and the statistical variations in the number of ions exposing each pixel (shot noise). The latter two limitations will be discussed in Sections IV.D and V.B, respectively.

A number of organic resists, both negative and positive, have been investigated to determine their applicability to IBL. Table II lists the proton and electron sensitivities of several resists compiled by Jensen [41a]. While H^+ ions have characteristics that promise to be most advantageous for high-resolution lithography with organic resists, there has been considerable interest in the sensitivity and range of heavier ions interacting with these resists. Again, Jensen [41a] has compiled these data, and they are reproduced in Tables III and IV, giving the effect of ion mass on sensitivity and penetration, respectively. Given the range and sensitivity of ions with mass and energy, it would be possible to utilize different ions at different energies for specific applications and processes.

The LMI sources that will be discussed in Section IV.C provide the means of obtaining focused beams of a variety of ion species with adequate current densities and submicrometer resolution.

TABLE II
Proton and Electron Sensitivities of Selected Resists[a,b]

Resists	$\overline{M}_w(\times 10^{-5})^c$	Sensitivity			Reference
		E (keV)	H^+/cm^2	e^-/cm^2	
Negative					
Polystyrene	1.6	1500	3.7×10^{13}	3.3×10^{14}	42a
	3.2	100	2.4×10^{12}	2.9×10^{14}	42b
Poly(glycidyl methacrylate cochlorostyrene)	2.0	1500	2.6×10^{12}	1.4×10^{13}	42a
Novolac	0.07	1500	9.0×10^{14}	1.0×10^{16}	42a
AZ1350J	—	180	1.9×10^{13}	6.2×10^{13}	43a
Poly(glycidyl methacrylate coethylacrylate)	—	100	3.4×10^{11}	2.6×10^{12}	42b
Poly(4-chlorostyrene)	2.9	100	1.3×10^{12}	2.9×10^{13}	42b
Poly(4-bromostyrene)	5.3	100	6.6×10^{11}	1.1×10^{13}	42b
Iodinated polystyrene	5.5	180	1.9×10^{12}	7.9×10^{12}	43a
Poly(chloromethyl-styrene)	1.2	100	9.0×10^{11}	1.3×10^{13}	42b
Poly(methyl methac-rylate)	8.0	1500	1.3×10^{13}	1.2×10^{14}	42a
	9.5	1500	3.9×10^{13}	—	42c
	6.0	180	1.0×10^{13}	—	43b
	—	150	2.0×10^{13}	—	19b
	4.0	100	2.6×10^{13}	1.2×10^{15}	42b
	—	50	1.3×10^{13}	—	43c
Poly(butenesulfone)	20	1500	3.1×10^{12}	4.7×10^{12}	42a
Poly(tetrafluoropropyl methacrylate) FPM	—	50	1.9×10^{12}	—	43c
Poly(methyl α-chloro acrylate)	15.5	100	1.8×10^{12}	2.4×10^{14}	42b
Poly(trifluoroethyl α-chloroacrylate)	2.7	100	3.2×10^{12}	6.0×10^{12}	41a

[a] From J. E. Jensen, *Solid State Technol.* **27**, 147 (1984).
[b] Electron exposures were at 20 keV; resist thicknesses 0.5–1.0 μm.
[c] Molecular weight data are the same for both H^+ and electron exposures and is the weight-average molecular weight.

Several patterning methods and types of resists have been proposed and investigated, for example, ion-implanted resists, ion-activated selenide resists, and self-developing resists. Silicon ions (among others) can be implanted in organic resists, making the resists strongly resistant to reactive ion etching [44a]. A focused ion beam or a projected ion image could be used to pattern the resist, so a protective etch mask is produced and in the

TABLE III

Effect of Ion Mass on Sensitivity of Resists[a]

Resist	Ref.	Ion	E (keV)	Sensitivity (ions/cm²)
Polystyrene	42 a	O^+	1500	4.3×10^{11}
		He^+	1500	1.4×10^{12}
		H^+	1500	3.7×10^{13}
		e^-	20	3.3×10^{14}
PMMA	35	Ar^+	120	2.0×10^{11}
		He^+	120	5.2×10^{11}
		H^+	120	3.6×10^{12}
		e^-	20	3.0×10^{14}

[a] From J. E. Jensen, *Solid State Technol.* **27**, 147 (1984).

TABLE IV

Penetration Range of Ions
in Organic Resists[a,b]

Ion	E (keV)	Range (μm)	Reference
Ga^+	40	0.046	35
	55	0.06	41c
	120	0.12	35
Ar^+	120	0.2	35
Si^{2+}	140	0.35	41b
O^+	1500	3[c]	42a
B^+	100	0.58	33
Be^+	100	0.75	33
Li^+	100	0.85	33
He^+	40	0.44	35
	100	0.92	33
	120	0.96	35
	1500	8[c]	42a
H^+	40	0.52	35
	100	1.08	33
	120	1.12	35
	240	1.85	35
	1500	40[c]	42a

[a] From J. E. Jensen, *Solid State Technol.* **27**, 147 (1984).

[b] Range values are estimated from experimental results unless otherwise indicated.

[c] Calculated range.

subsequent ion etching process the equivalent of a negative resist obtains. High doses are required ($1-20 \times 10^{15}$ ions per square centimeter) to provide adequate resistance to the etching process, but at least in the case of ion projection lithography if a satisfactory ion beam is available, this procedure would not be too slow.

Inorganic resists have also been investigated. Germanium selenide (GeSe) with a thin layer of silver (100 Å) can be exposed to low-energy ions (e.g., He^+, N^+, Ar^+, Xe^+, and Ga^+). The ions cause the migration of silver into the GeSe in the exposed areas, producing a negative resist image [44b]. Again sensitivity is relatively low ($\simeq 5 \times 10^{13}$ to 10^{15} ions per square centimeter). However, the sensitivity can be improved if the GeSe is deposited at an oblique angle ($\simeq 2 \times 10^{12}$ N^+ per square centimeter). Other inorganic resists such as the halides NaCl, LiF, and AlF_3 have been investigated as possible very high resolution resists patterned by nanometer-diameter, high-current-density electron probes obtained with the STEM [44c]. While very high doses are required, the higher sensitivity available with ion beams and the fact that these resists are self-developing make this type of resist one that should be investigated for IBL.

Another self-developing resist that has been investigated for ion-beam exposure is nitrocellulose [44d]. Thin layers of this organic material completely decompose, leaving little or no residue when exposed to adequate doses of radiation.

IV. ION-BEAM LITHOGRAPHY SYSTEMS AND INSTRUMENTATION

The methods and instrumentation that are used for IBL depend primarily on the characteristics of the ion sources available. Until very high brightness field-ion sources were developed, IBL was limited to exposure of resists and substrates by proximity printing through stencil masks using duoplasmatron or similar ion sources [18,45]. The important advance that has made it possible to do IBL by direct writing with focused ion beams has been the development of field ionization sources. These sources have high brightnesses ($10^6 - 10^8$ A/cm^2 sr) and produce ion beams with high angular current densities and low energy broadening. By means of well-developed charged-particle optics these beams can be focused to high-resolution, high-current-density probes that can be scanned to expose substrates in the desired patterns for lithographic processing. Thus, IBL using focused probes is quite analogous to EBL, and very similar instrumentation and systems can be used.

A basic IBL system will be described briefly here to provide an overview

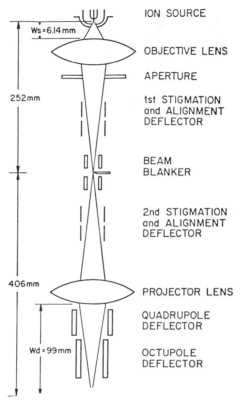

Fig. 13. Schematic of an electrostatic optical system for a direct write IBL instrument using an H_2^+ field-ion source. [From H. Paik, G. Lewis, and B. M. Siegel, *J. Vac. Sci. Technol.*, B **3**, 75 (1985).]

of the instrumentation before the several components are discussed in detail. Figure 13 is a schematic of a typical optical configuration for an instrument that is a focused-ion-beam lithography system using a field ionization source [46]. The characteristics of the source determine the design of the ion optics and the limits of the capabilities of the IBL system. In addition to the source, the system column consists of the ion lenses, the alignment deflectors and stigmator, the beam blanking, the scanning deflection section, and the laser interferometer controlled stage on which the substrate is mounted.

Given the characteristics of field ionization sources, some general considerations must be taken into account in designing an ion optical system that will produce the desired ion probe at the substrate plane and allow the probe to be scanned over the desired area. In general the first lens focuses the source to a crossover at a blanking aperture that is placed midway between the blanking deflector plates. Since the effective source size is very small in field-ion sources and the wavelength of the ions is so small that diffraction effects are negligible, the axial image figure is determined by the chromatic and spherical aberrations, usually of the first lens. To obtain the highest-resolution probe with the largest current density this lens is carefully designed to have low chromatic and spherical aberration coefficients and a short focal length. The lens is placed as close to the source as possible to obtain a large acceptance angle of the ion beam from the source. With the first lens adjusted to focus the source to an image at the blanking aperture, the crossover at the blanking aperture is then focused by the projector lens on the substrate. A lens of relatively long focal length is needed for projector lenses to give the long working distances desired in IBL optical systems and thus minimize the deflection aberrations. The deflection aberrations will determine the image figure as the beam is scanned to the peripherial areas of the field and set the size of the field that can be exposed without stepping the stage. The stage position, monitored and under the control of laser interferometers, will have to be of the highest accuracy if the full potential of IBL is to be realized. IBL is still in the process of experimental development, so the stages that have been developed for EBL systems have been used on most instruments that have been constructed. Figure 14 is a schematic of the optical system of the IBL system produced by VG Semicon Ltd., which uses an LMI source. This commercially available instrument produces ion probes of <0.1 μm diameter with current densities of gallium of 5 A/cm^2 over scan fields of 1 × 1 mm at 90 kV.

A. Field-Ion Sources

The characteristics of the source are the determining factors in the design of the ion optical system, so a detailed discussion of the ion sources used in IBL systems will be presented first. It is important to realize that these sources are still in the early developmental stages and their characteristics are constantly being enhanced.

The first type of field-ion sources to be developed with angular beam currents and energy spreads that were adequate to be focused to high-resolution probes were LMI sources [3,4,10–12]. More recently, gaseous field ionization sources (GIFSs) have been developed to a stage where they have the potential of giving very high resolution, very high current density probes [13].

The LMI sources have a variety of applications in the processes involved

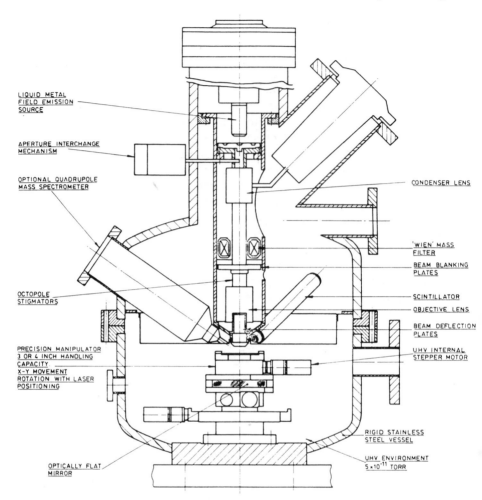

Fig. 14. Schematic cross section of the column of the VG Semicon focused ion beam system. Model IBL-100S using an LMI source. VG. Semicon, Ltd. East Grinstead, England.

in the microfabrication of microelectronic devices [18]. While, as already pointed out, LMI beams are not satisfactory for direct patterning in resists by focused ion beams, they are of great interest in the microfabrication process, and there is a great deal of work being done on the development and application of these sources. For example, among the most important applications is the possibility of doping semiconductors by direct implantation (without masks) of the specific ion species required in device materials.

On the other hand, GIFSs are higher brightness sources than LMI sources and can produce a beam of light ions that can be focused to a very high resolution, high-current-density probe that can be used to pattern resists directly. As far as the application of GFISs are concerned, they are still very much in the development stage. At this point, only hydrogen and helium and some of the rare gases have been investigated as sources. However, there is the potential of using other gases in a variety of applications. There is an extensive literature on gaseous field ionization, for this source is based on the field ion microscope (FIM) invented by Mueller [47a]. The FIM is a powerful tool for studying crystal structure, defects in solids, radiation damage, etc., at the atomic resolution level, and studies

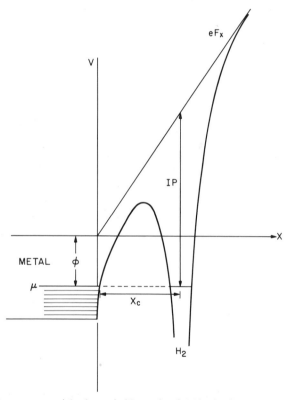

Fig. 15. Electron potential schematic illustrating field ionization of H_2^+ from a metal tip. Here, V is the electron potential as a function of distance x from the tip, μ the Fermi level, ϕ the work function, IP the ionization potential of the H_2 molecule, F the applied field, and x_c the critical distance at which there is a high probability of an electron from the hydrogen molecule tunneling into the metal (at the Fermi level), leaving an H_2^+ ion that is accelerated by the field F_x.

have been made both on the physical processes involved in gaseous field
ionization in the FIM and on crystal structure and other applications of the
FIM [47a–d]. Since the basic physics that underlies the gaseous field
ionization process has been developed and reported in the literature on the
FIM, only a brief description of the relevant physics will be presented in
this chapter.

B. Gaseous Field-Ion Sources

Gaseous field ionization processes can be understood by reference to Fig.
15, which is an electron potential diagram illustrating the physical process

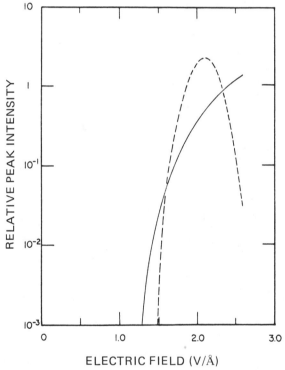

ELECTRIC FIELD (V/Å)

Fig. 16. Ratios H^+/H_2^+ (——) and H_3^+/H_2^+ (---) versus applied field. The plot shows ratios
corresponding to peak heights rather than total ion current. Since the peaks of all species are
narrow at low fields, these ratios approximate total ion production ratios at low fields. At
higher fields the H^+ distribution broadens relative to H_2^+, while H_2^+ stays sharp. Therefore, the
ratio H_3^+/H_2^+ for the total ion production decreases, and the ratio H^+/H_2^+ increases more
rapidly with increasing field than the plots shown here imply. [From A. Jason, B. Halpern, M.
Inghram, and R. Gomer, *J. Chem. Phys.* **52**, 2227 (1970).]

that occurs when an H_2 molecule approaches a metal surface (the emitter tip) that is at a high positive potential. When the molecule comes to a critical distance x_c from the metal, where the energy level of the electrons in the molecule are at the Fermi level of the metal, there is a high probability that the electron will tunnel into the metal, leaving an H_2^+ ion in the high-field region so the ion is then accelerated. Since the field is produced by a gradient of the potential around the tip, which has an approximately spherical surface, the ions have radial trajectories out from a virtual point in the tip. The ionization occurs over a very narrow distance from the tip, for if the molecule is closer to the metal than x_c, the energy of the electron in the H_2 molecule is lower than the Fermi level in the metal and there is no vacant state for the electron to tunnel into; and if the molecule is too far from x_c, the probability of tunneling through the barrier is very low. If the field at the tip is kept below ~ 1.5 V/Å, there will be negligible secondary ionization or dissociation processes to produce H^+ ions. If a field of ~ 2 V/Å is applied, the H_2^+ will also be dissociated to give an ion beam that is a mixture of H_2^+ and H^+. There is also a region in which H_3^+ ions are produced. These phenomena are illustrated in Fig. 16 taken from the investigations of Jason et al. [48]. There is a voltage termed the best image voltage (BIV) at which a high-resolution image of the atomic structure of the emitter is obtained. Such a field-ion micrograph of a $\langle 110 \rangle$-oriented tungsten tip is shown in Fig. 17a. In the case illustrated here the tip has been field-evaporated by applying a field high enough to strip off atomic layers of tungsten, so imaged in this figure are the individual atomic sites of the crystal planes of this tungsten tip.

To obtain a beam from a GFIS that can be focused to a high-current-density probe and adequately meet the requirements for IBL, the source characteristics must meet some stringent conditions. The source must produce a beam of at least the order of nanoamperes with high angular confinement ($\geq 10 \, \mu A/sr$) from a single emission site that has high stability and is located on or near the optic axis of the system. To achieve these characteristics the emitter structure must be (1) processed to have the desired localized site, (2) configured to provide a high gas supply to the emitter, and (3) operated at the optimum electric field and the optimum temperature of both the emitter and gas supply.

Some early attempts were made to use GFISs in probe-forming systems. In one of the first applications, Levi-Setti and co-workers [49a,b] used a field-ion source to produce a hydrogen ion probe for a scanning transmission ion microscope. They were able to focus the ion beam to a high-resolution probe (≈ 50 Å) but with a very low current ($10^5 - 10^3$ ions per second) of mixed H_2^+ and H^+ ions, making it quite impractical for IBL. Another ion-beam system designed to write in resists with a scanned

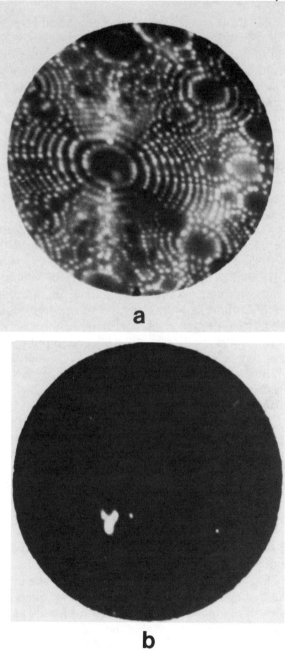

Fig. 17. (a) Field-ion microscope image of a field-evaporated W(110) tip at best image voltage. (b) Confined H_2^+ ion beam from a single site on an annealed (110)W tip. [From G. R. Hanson and B. M. Siegel, *J. Vac. Sci. Technol.* **16**, 1875 (1979).]

hydrogen ion beam was developed by Orloff and Swanson [49c,d]. They improved the source by using differential pumping to allow an increased pressure of hydrogen in the region around the emitter tip and obtained angular beam currents of 0.5 to 1 μA/sr, still marginally useful for obtaining an ion beam that could be focused to a probe for writing in resists. Since they operated the source at fields of 2.3 to 2.6 V/Å the ion beam also had a mixture of H_2^+ and H^+ with an energy spread of some 3 eV. They were able to focus the beam to a submicrometer probe, but their image was severely limited by chromatic aberration.

A very high brightness H_2^+ ion source with significantly better characteristics was developed by Hanson and Siegel [13] by modifying the GFIS system to operate under conditions that gave emission from a single site, had a high-supply function to the emitter, and gave a single ion species, H_2^+, with a low energy spread. The modifications that produced these improvements included cooling the emitter tip to liquid-helium temperatures, operating and processing the tip in an ultrahigh-vacuum environment to produce single-site emission, and adjusting the ionization field to 1 to 1.5 V/Å so only the parent ion, H_2^+, is produced. A columnated beam was obtained with typical angular currents of 10 to 20 μA/sr and an energy spread of 1.0 eV (FWHM) [50]. A schematic of this source is shown in Fig. 18. The tip was cooled by mounting on a sapphire rod that was clamped to a liquid-helium cold finger. The sapphire is an excellent thermal conductor at low temperatures but is also a very good electrical insulator, so the cold tip could be raised to a high electrical potential. The tip was etched to a radius of some 1000 Å, so a voltage of 4 to 6 kV produced a field of 1.0 to 1.5 V/Å at the tip. A cathode structure surrounded the tip with only a 1-mm-diameter aperture for the beam to exit and to provide the small orifice required for differential pumping. The H_2 gas was cooled and led into the volume surrounding the tip, so a pressurre of $\simeq 10^{-3}$ torr could be maintained in that region while the differential pumping in the system outside the cathode produced an ambient gas pressure of only 10^{-6} to 10^{-7} torr in that region where the electrostatic ion optical system accelerated and focused the beam. The image of this beam obtained in the early experiments is shown in Fig. 17b. This beam, emitted from an apparent defect in the thermally annealed tip, had a very high angular current density of 60 μA/sr. The processing of the tip to produce a structure that has a single site at which field emission occurs is a critical step in obtaining beams with high angular current. To obtain a practical source the emitter structure must also have long-term stability and have the emission site located so as to produce a confined ion beam that is close to the optical axis of the system. Successful tip processing is crucial to producing a useful GFIS.

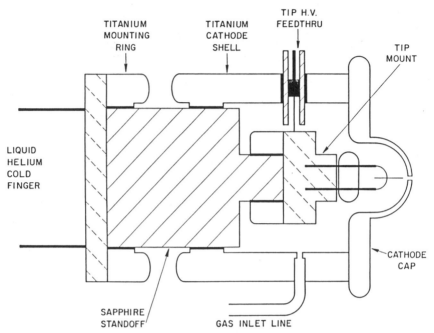

Fig. 18. Schematic of a gaseous (H_2^+) field-ion source. [From R. J. Blackwell, J. A. Kubby, G. N. Lewis, and B. M. Siegel, *J. Vac. Sci. Technol.*, B **3**, 83 (1985).]

1. Emitter Configuration and Processing

The emitter in the initial source of Hanson and Siegel [13] was a $\langle 110 \rangle$-oriented tungsten tip that was annealed to produce a smooth surface over which the physisorbed H_2 diffused to a few random sites at which the field was high enough, ≥ 1 V/Å, to cause tunneling of an electron from an H_2 molecule near the surface into the metal with the consequent ionization and acceleration of the H_2^+. A typical annealed tip would have a few bright sites that competed for the hydrogen supply. Usually there was a dominant site that gave very stable emission for periods of hours, but eventually a new emission site appeared, caused, most likely, by an atomic rearrangement induced by a contaminating molecule such as N_2 or O_2. Since these events and the location of the emission sites were random and were of atomic dimensions, they had limited stability. The emitter tip processing had to be improved to obtain larger, stable sites that could be produced at the apex of the tip to give a well-defined, oriented beam with high angular current density.

Three approaches have been tried. Brady [51] used a ⟨100⟩-oriented single-crystal tungsten tip that was "built up" by an oxygen-enhanced thermal-field process [52a,b;53], causing a faceting of the tip to produce a tip that gives good angular confinement of the electron beam when it is used in electron field emission. When properly "built up" high angular electron-beam confinement is achieved, reducing the divergence of the electron beam to a half-angle of ∼ 0.15 rad. To obtain an emitter configuration that is better for gaseous field ionization, the processing of the tip is continued by further thermal treatment with the field reversed. A single emission site is formed at the apex of the tip, and an axially oriented ion beam is obtained. However, the emission site is composed of a relatively few atoms and therefore does not have the stability desired for long-term lithographic processing. This tip-processing method has been used to obtain an ion source that can be focused to an H_2^+ probe for experimental exposures of resists [54].

The two other methods of processing emitter tips to obtain stable, reproducible emitters that are under development have produced interesting and very promising results. Schwoebel and Hanson [55] reported a method for producing a stable "microboss" on annealed ⟨100⟩-oriented tungsten tips within ± 10° of the apex. Their initial procedure involve electron field emission through a condensed hydrogen film on the tip cooled to ≃ 5 K. In the process hydrogen ions are produced that bombard the negatively biased tip and cause a rearrangement of the tungsten atoms. A structure is nucleated and is allowed to grow until it is estimated to be some 200 Å in diameter and 30 Å high. This growth process is observed by monitoring the field emission microscope pattern and the electron-beam current. When the desired microboss has formed, the polarity of the tip is reversed and an H_2^+ beam is obtained that typically has a current of 10 to 15 nA confined within a 16-mrad half-angle. The long-term stability of these structures is enhanced by an anneal at 1100 K in the presence of a positive field. The supply function to the emission site and the ionization voltage at which this tip operates can be controlled by choosing the radius of the annealed tip [56a]. As Hanson pointed out [56a] a large-radius tip with a small microboss provides a large high-field area on which gas is adsorbed and thus gives a high supply function. The large-radius tip requires a high voltage (≃ 50 kV) to produce the required field of 1.0 to 1.5 V/Å at the microboss, and the ion optical system will have to be designed to accommodate this situation. Schwoeble and Hanson reported additional details on the characteristics of their GFIS obtaining both H_2^+ and He^+ ion beams [56b].

Schwoebel [56c] has published the most recent modifications of the methods used to configure microbosses on emitter tip by ion bombardment. They now use single-crystal iridium wire that is etched to a point, annealed in oxygen at a pressure of $\simeq 10^{-6}$ torr to remove carbon, and subjected to bombardment by helium ions in a manner similar to the hydrogen ion bombardment of tungsten tips. He found that hydrogen caused embrittlement of tungsten and the microbosses grown on the tungsten would be destroyed by the high fields. The microbosses produced on iridium by He^+ bombardment are much more stable, particularly if they were grown with the iridium tip kept at elevated temperatures (up to $\simeq 700°C$).

Another method for producing a stable, oriented emitter structure has been developed by Kubby and Siegel [57a]. In this method a tungsten tip is ion-milled while it is rotated around the axis of the tip. The surface is configured by the morphological changes that occur with the erosion by sputtering under ion bombardment. On the microscopic scale (100–1000 nm) where the erosion is on a scale that is large compared with the ion penetration depth, a conical surface is obtained that has an included angle determined by the ion-beam energy and angle of incidence on the tip. Also at the apex of the cone there will be an erosional sharpening on a scale that is related to the penetration depth of the incident ions (10 to 100 nm) and a rounding of the sharpened tip by surface migration of the tungsten [57b]. A spherical protuberance some 10–20 nm in diameter and 10 nm high is

50 nm

Fig. 19.   Tungsten emitter tip for gaseous field-ion source configured by ion sputtering. [From J. Kubby and B. M. Siegel, *J. Vac. Sci. Technol. B* **4**, 120 (1986).]

produced and thus locates the emission site for field ionization on the axis of the tip. A shadow transmission electron micrograph of such an ion-milled tip is shown in Fig. 19. The morphology of this tip is the one required for a satisfactory GFIS; it has an emission area located at the apex of the tip; the site is made up of many atoms and therefore produces a stable beam since the site will not be affected by small atomic rearrangements if contaminant molecules impinge on the tip.

2. Gas Supply

The other requirement for producing a high-brightness ion source with adequate beam current confined within a small angular divergence is the provision of a large supply of the desired molecular or atomic species to the emission site. The supply of gas in the emitter region is enhanced by the polarization of the gas molecules in the high-field region of the tip. From kinetic theory, the normal value of the supply from the gas phase without field effects is given by

$$S_0 = p/\sqrt{2\pi MkT}, \tag{11}$$

where p is the ambient gas pressure, M the molecular weight of the molecule, T the temperature, and k Boltzman's constant. The polarization force, in a field $\mathbf{F}$

$$\alpha_p(\mathbf{F} \cdot \nabla)\mathbf{F},$$

where α_p is the polarizability of the gas molecule, enhances the supply S by a factor σ, so

$$S = \sigma S_0 \simeq \frac{1}{4}\left[\frac{\alpha F^2}{2kT} + 2.7\left(\frac{\alpha F^2}{2kT}\right)^{2/3} + 2.7\left(\frac{\alpha F^2}{2kT}\right)^{1/3} + 1\right]S_0, \tag{12}$$

according to van Eekelen [58], who based his derivation on a hypertoloidal field approximation.

At low temperatures the gas molecules can be accommodated or physisorbed on the tip surface for a mean finite lifetime τ and supply the emitter site by surface diffusion. The supply function then becomes enhanced by the area of the tip on which the gas is accommodating and has a high probability of being transported to the emission site. The radius of this area is given by the surface supply length $L = (2D\tau)^{1/2}$, where D is the diffusion coefficient of the molecules accommodated on the surface.

When the emitter site is supplied from the gas phase, Eq. (11) indicates that the ion current will be directly proportional to the ambient gas pressure p. As the tip is cooled to the temperature at which the supply is surface diffusion limited, a $\sqrt{p}$ dependence will be observed. In any case the tip configuration desired is one in which there is as large a region of high field as possible, consistent with the operating voltages set by other instrumental factors. Supply to the emitter site will also be strongly enhanced by lowering the tip and gas temperatures to the point at which the gas freezes out or does not have a high enough surface diffusion rate. In the case of the original H_2^+ source developed by Hanson and Siegel [13,50], in which the emitter tip was cooled by a liquid-helium cold finger to ~ 15 K, the hydrogen gas is physisorbed on the surface of the tip and supplies the emission site by diffusion from an apparent area ~ 400 Å in diameter as well as directly from the gas phase in the high-field region near the tip. A pressure dependence of approximately $p^{2/3}$ was observed.

3. Field Ionization of Hydrogen

As pointed out earlier, to obtain an ion beam with only the single species H_2^+, the field at the emitter must be kept below $\simeq 1.5$ V/Å. When the field is

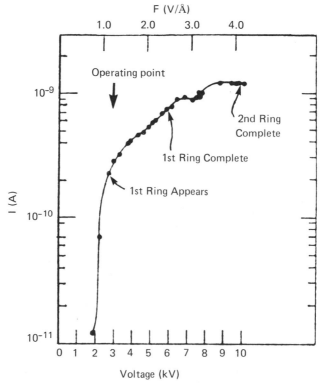

Fig. 20. Plot of total ion current (in sealed field-emission tube) versus voltage. [From A. Jason, B. Halpern, M. Inghram, and R. Gomer, *J. Chem. Phys.* **52**, 2227 (1970).]

increased above that value secondary ions, H^+ and H_3^+, are produced by dissociation and combination [48,58]. The plots in Fig. 16 show the relative amounts of H^+ and H_3^+ compared with the parent ion H_2^+ as a function of the field from the data given by Jason *et al.* [48]. By plotting the ion-beam current as a function of voltage or field for a given-radius emitter tip, Jason *et al.* obtained the curve shown in Fig. 20 (see also van Eekelen [58]). As long as the field is kept below $\simeq 1.5$ V/Å, the H_2^+ beam current rises very steeply. There is a "knee" in the curve at this point, and the secondary effects set in. From the field at the "knee" and the data in Fig. 16 it can be seen that the amount of H^+ and H_3^+ at the "knee" will be insignificant, and this is the point at which the emitter should be operated to obtain the largest H_2^+ ion current densities when beams are focused by lenses for $\Delta E \simeq 1.0$ eV at the knee [50]. At fields above the "knee" there are mixtures of H^+ and H_3^+ ions with large energy spreads [48].

As the field is increased above the "knee," the beam is also less confined. The field-ion micrographs in Fig. 21 show the change in the ion emission patterns from a thermal-field-processed emitter tip [51]. The greatest beam confinement was obtained at 4.2 kV for this emitter. As the voltage was increased, other regions of the tip were above the critical fields and the ionization sites spread out, finally to ring structures where surface accommodation and diffusion determined the emission pattern. Clearly, only a beam with a high degree of confinement as shown in Fig. 21c can be focused to a high-resolution probe.

4. Characteristics of the H_2^+ Field-Ion Source

The gaseous field ionization source is a very high brightness source that holds great promise for IBL. The size of the virtual source in the H_2^+ sources that have been developed is very small, of the order of 5 to 10 Å in diameter, so with an angular current of 20 μA/sr, a brightness of $\simeq 2 \times 10^9$ A/cm^2 sr is obtained for the "virtual" source with the ions accelerated to 6 kV in the ionization process. However, in any practical application, this beam must be focused, and since electrostatic lenses with their relatively high aberration coefficients must be used, the image figure of the focused probe will be chromatic or spherical aberration limited (or both). A beam with an angular current of 20 μA/sr focused to a 100-Å-diameter probe and accelerated to 50 kV would have a brightness of $\simeq 7 \times 10^7$ A/cm^2 sr. This is, of course, a very high brightness beam, comparable to field electron emission sources and considerably better than the brightness that can be achieved in focused probes of heavy ions using LMI sources. The poorer probe characteristics obtained with beams focused from LMI sources are a consequence of the larger energy spread in the beams (5–20 eV) and their

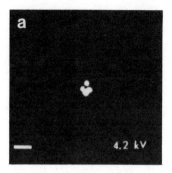

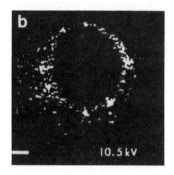

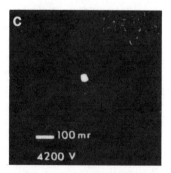

Fig. 21. H_2^+ ion patterns from a thermal-field-processed $\langle 100 \rangle$ tungsten tip. (a) Beam confinement at field-ion voltage of -4.2 kV, $F = 1.5$ V/Å, $dI/d\Omega = 10$ μA/sr. (b) Pattern with voltage at -10.5 kV. (c) Final beam confinement with voltage lowered to -4.2 kV, $I_b = 1.5$ nA. (White lines subtend 100 mrad.) [From J. R. Brady, M.Sc. Thesis, Cornell Univ., Ithaca, New York, 1981.]

TABLE V

Characteristics of Ion Sources[a]

Characteristic	Duoplasmatron	Liquid metal ion source (see Table VI)	Gaseous field-ion source
Ion species	H^+, He^+, Ne^+, Ar^+, Xe^+, N^+	Ga^+, In^+, Bi^+, Au^+, Be^+, Be^{2+}, B^+, Al^+, Si^+, Si^{2+}, Ge^+, As^+, As^{2+}, Sn^+, Sb^+, Sb_2^+	H_2^+, He^+
Angular current density	> 100 mA/sr	$1-40$ μA/sr	$1-20$ μA/sr
Virtual source diameter	50 μm	$40-67$ Å	$5-10$ Å
Brightness			
Virtual source (A/cm² sr)	$> 10^4$	$> 10^6$	$> 10^8$
Focused probe at 50 keV	—	$> 10^5$	$> 10^7$
Total emission current	> 200 μA	$\simeq 1-100$ μA	$1-100$ nA
Typical beam current	> 200 μA	$0.1-20$ nA	$0.1-20$ nA
Energy width (eV)	± 10 V	$5-40$ V	1.0 (FWHM)

[a] References: duoplasmatron [18,98a,b,99]. Liquid metal ion source [10,12,22a,b,59]. Gaseous field-ion source [13,50].

larger virtual source size (40–70 Å) [59]. Table V lists characteristics of these two types of sources.

There are, of course, complexities and limitations in the GFISs at their present state of development. Only experimental prototype sources have been developed, and they require instrumentation that challenges the most advanced procedures in experimental physics and current technology. A number of inherently incompatible conditions are required: The emitter tip must be cooled to cryogenic temperatures and isolated from heat loads to it, the tip structure must also be electrically insulated so it can be floated to the voltage to which the ion beam is to be accelerated, and the whole system must be in a very clean ultrahigh vacuum so the emitter tip can be thermally processed and its operation not affected by contaminating gas molecules. Thus, a rather formidable set of conditions is required to obtain a satisfactory high-brightness H_2^+ field-ion source, and these conditions have so far been realized only in a laboratory environment.

However, since the experimental results with the H_2^+ source have demonstrated the very superior performance to be obtained with this source and the required methods are all available in the current technology, efforts are being made and will continue to be made to develop this source for operation in a production environment. GFISs are in the earliest stages of development. There is the possibility of producing very high brightness sources of other ion species using other gases, either the rare gases (helium, argon, neon, etc.), reactive ions such as oxygen and nitrogen, or a variety of species using molecular gases that would produce beams of the desired species by field dissociation. These ion beams could have important applications such as doping, enhanced reactive ion etching, mask repair, and the other nonlithographic applications of focused ion beams in the submicrometer fabrication processes discussed earlier. The complexity of the instrumentation involved in developing a GFIS that produces a confined ion beam from a single emission site has perhaps discouraged many workers from pursuing this task. However, research and development at Cornell is continuing on GFISs, and other gases as well as hydrogen are being investigated. Some Japanese investigators have reported that they are also working on gaseous sources [60].

C. Liquid Metal Ion Sources

LMI sources are the other type of high-brightness ion source that can be focused to probes of submicrometer dimensions with adequate current densities for direct write processing in microfabrication. This type of ion source was first suggested by Krohn [61a,b] for ion propulsion of spacecraft. Clampitt and co-workers [3,4] at Culham in England developed this source to produce a variety of species of ion beams. The potential of these ion sources for microfabrication processes was recognized, and several investigators began to study the LMI source as well as apply it to ion-beam processing [5a,b,10-12] at that time. The interest in this source has developed to the degree where special sessions devoted to various aspects of the investigation of these sources and their application to microelectronic processing are held at meetings and symposia. The annual *International Symposium on Electron, Ion, and Photon Beams* now includes several papers on this subject, and these papers are published in an issue of the *Journal of Vacuum Science and Technology* each year. Sessions and symposia on this subject have also become part of the International Symposium on Field Emission, with the proceedings published in special issues of the *Journal de Physique*. The reader is referred to these journals for an update on the work being carried out in this field (see Refs. [62] and [63]).

An ion beam can be produced by an LMI source using any metal that has a relatively low melting temperature and low vapor pressure. LMI sources have been developed to a stage where they can give reliable, steady ion beams of a wide variety of ion species. These sources are easy to fabricate and operate under conditions that can be realized in a prototype production environment, so a number of instruments have been constructed that can produce focused ion probes of several different ion species, and some ion-beam systems are commercially available. LMI ion sources with lenses that produce focused submicrometer probes with good current density ($1-5$ A/cm^2) are also available commercially [64a,b].

1. Applications of Ion Beams from LMI Sources

The large interest in LMI sources has been generated by the practical applications that the availability of focused beams of a variety of ion species from these sources makes possible. Some of the systems that have been built and reported in the literature will be described below. Perhaps the most interesting applications are the high-resolution doping of semiconductor materials in fabricating microelectronic devices. Since the probes can be focused to diameters of <0.1 μm, it is in principle possible to implant doping elements with desired profiles [19a,b]. The possibility of tailoring the doping at this high level of resolution provides an extremely important tool for experimental development of devices. There has also been an interest in direct structuring at high resolution using these ion beams for sputtering the substrate [19b,59a,b] or ion-beam-assisted reactive ion etching [20,22a,b]. Another application has been the repair of metal masks with these focused ion beams, either sputtering out undesired structures or repairing the masks by ion deposition in holes or similar faults [7]. While most of these procedures should not strictly be called lithography, all of them represent very important processes in the microfabrication of microelectronic devices.

The beams of heavy metal ions from LMI sources have been used in the direct writing of patterns in resist, but these procedures do not appear to have much promise. The straggling of the heavy ions in the resist caused by ion–nuclear interactions produces an undesirable profile in the developed resist, as was discussed in the section on ion interactions with substrates. Since this straggling produces a "proximity effect," there do not seem to be any advantages of heavy-metal ILB over EBL. There may be some advantage if multilayer resists or resists such as AgGeSe are used. In any case, since the instrumentation of focused-ion-beam systems having LMI sources and their methods of application are the same as or very similar to focused probe systems using charged particles in patterning for lithography, these systems will be discussed in this chapter.

2. Characteristics of LMI Sources

While the characteristics of many LMI sources have been investigated and described in the literature, the physical mechanisms of the operation of

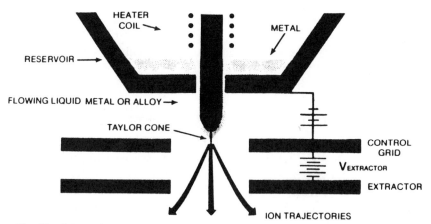

Fig. 22. Schematic of a liquid metal ion source. Courtesy Ion Beam Technologies, Inc. Beverly, Massachusetts.

these sources are not well understood, and a number of characteristics have not been explained. A schematic of a typical LMI source is shown in Fig. 22. One of the first LMI sources to be developed was the Ga$^+$ source. This source has been studied extensively because gallium metal is particularly well suited for producing an ion beam from a liquid metal [5,10]. Gallium has a melting point of 30°C with a vapor pressure of $< 10^{-12}$ torr and can be made to wet a tungsten tip if properly treated. If a voltage of about -16 kV with respect to a tip is applied to the cathode, a field of some 2.0 V/Å is produced at the tip. The molten gallium is pulled into a cone by the stress of the field until the surface tension of the liquid with decreasing radius of curvature is equal to the field stress. Taylor [65], who was the first to study this phenomenon using water as the liquid, calculated an equilibrium configuration of a conical shape with an included angle of 98.0° (the "Taylor cone"). Recent studies [66,67], both theoretical and experimental, have discussed modifications of the Taylor cone model, particularly when applied to actual ion emission from conducting liquids. When a critical field strength is reached, a small overvoltage causes the gallium to come off the tip as a beam of Ga$^+$ ions. In the case of Ga$^+$ beams angular currents of 20 to 60 μA/sr have been observed with energy spreads starting at 4.5 to 5.0 eV at the lowest currents and increasing to some 20 eV at the higher currents [10]. Any metal or material that can be held on the tip (or in some cases in a small capillary tube) in the liquid phase and has a low vapor pressure can be used to produce an ion beam in the same manner as described for gallium.

   Among the first studies were those employing ion beams of gold, indium, bismuth, and cesium [3,4,10]. An important development was made when alloys were used to obtain species of ions that did not have pure elements that could meet the requirements of a reasonable melting temperature and low vapor pressure. Gold alloys of silicon and platinum and palladium alloys of boron and arsenic were among the first to be investigated [5b]. Separation of the different ions was then necessary, and some type of mass separator had to be employed. Seliger and co-workers [5b] used an $E \times B$ analyzer. Other laboratories followed, particularly in Japan, where extensive work is being done with LMI beam sources and systems. Namba and his group at Osaka University have been particularly active in this area [12]. Table VI lists the ion beams of different species, as well as characteristics of these sources that have been reported by Namba's group [12]. Since new results are being reported in ever increasing numbers, this table contains only a partial list of the alloys used and the ion species obtained.

The characteristics of these sources, the ion species available, and the quality of the beams they produce are of primary interest from the point of view of the microfabrication of microelectronic devices. However, it is important at least to discuss the physics of this system to understand the limitations of the LMI sources and be familiar with the questions on the mechanisms involved that are as yet unanswered. Certain characteristics of these sources place undesirable limitations on the resolution and current

TABLE VI

Characteristics of Liquid Metal Alloy Ion Sources[a]

Ion	Liquid metal	Current intensity[b] (μA/sr)	Emitter	Life (hr)
Be	Be–Au, Be–Si–Au	~20	W	>200
B	B–Pt, B–Ni–Pt, B–Ni	~10	W C	>10 200
Al	Al	~43	C BN–TiB$_2$	— —
Si	Si–Au, Si–Be–Au	~6	W	>60
Zn	Zn	—	W	—
Ga	Ga	~26	W	>200
Ge	Ge–Au	—	W	—
As	As–Pt, As–B–Ni–Pt	~4	W W	— —
Sn	Sn	—	—	—
Sb	Sb–Pb–Au	~1	Ni	>26
Au	Si–Au, Si–Be–Au	~24	W	>200

[a] From S. Namba, in "Electronic Device and Materials" (J. Chen, ed.), Taiwan, 1984.
[b] $\Delta E = 10$ eV.

densities that can be achieved in probe-forming systems. Among these is the relatively large energy spread of 5 to 40 eV [10], giving large chromatic aberration figures when the beam is focused with electrostatic lenses. Chromatic aberration thus limits the angular acceptance angle that can be used to achieve a desired resolution. As long as the image figure is chromatic aberration limited, the current density obtained in the focused probe will be constant over a range of probe diameters if the acceptance angle is scaled to obtain enhanced resolution. This condition follows from the fact that the diameter of a focused probe that is limited by chromatic aberration is linearly proportional to the angular acceptance angle of the optical system, so the area of the image is proportional to the square of the acceptance angle. Since the current in the beam in the focus probe also increases as the square of the acceptance angle, the current per unit area remains constant. Another characteristic of the beams produced by the LMI sources reported up to this time is the apparently large size of the virtual image of the source. Kumoro *et al.* [59] reported source diameters of 40 to 67 Å obtained by calculations based on the limiting performance of their probes.

3. Physical Model of the LMI Source

There has been considerable discussion of the actual physical mechanisms involved when the liquid metal source goes from the equilibrium condition to the emission of ions. While there have been some challanges [68] to the standard "Taylor cone" shape of the liquid metal in the LMI source, the consensus now is that the conical shape postulated by Taylor is the shape approached in the low current limit or at large distances from the emitting area [69,70]. However, the last two papers disagree significantly in their conclusions. The rounded Taylor cone could not reasonably explain the observed emission currents at the fields necessary to sustain field evaporation at the apex of the cone [66,71–73c].

Kingham and Swanson have developed the most complete theoretical model of LMI source operation yet presented in a series of papers [66a–c,67]. They conclude that field evaporation is the dominant current generating mechanism of singly charged monomer ions M^+ at the apex of the liquid surface on which current dependent protrusions occur and these protrusions sustain the field evaporation [66b]. These ions can be raised to higher positively charged states, M^{2+}, M^{3+}, or higher, by the high fields required for large ion currents. Field ionization of neutral atoms and clusters evaporated from the sides of the protrusion and pulled into the high field by polarization mechanisms also occurs but is not the dominant ion current mechanism according to the Kingham and Swanson model. Their model is consistent with the experimentally observed characteristics of LMI sources on energy spread, energy deficit and charge state ratios. They are also able to explain the stability observed in the currents by space charge effects that occur in their model.

While the physics of the field-emission processes involved in LMI sources has been developed to a reasonably satisfactory state, all the details are not completely understood. Active research is still going on to obtain the data we need to understand more of the mechanisms involved. In the meantime, LMI sources are very practical and are being used in a variety of instruments.

There are, of course, some characteristics of LMI sources that limit their application in the microfabrication processes. In particular, the relatively wide energy spread in these ion beams sets a chromatic aberration limit on the resolution and current density to which the probes can be focused. In addition, the ion species that are produced are either very heavy or at least much heavier than the substrate atoms. As discussed in the sections on ion–matter interactions this means that most of the energy of the incident ion is deposited by ion–nuclear interactions. As a consequence, there will be considerable straggling in resists (see Figs. 6, 11, and 12 for incident Ga^+ ions in PMMA), so in direct writing with a heavy-ion beam the resolution is poorer than direct writing with a H_2^+ ion probe.

The limitations on lithography with beams from LMI sources do not distract from the very important applications that can be realized with focused ion beams of selected species from LMI sources. These are related primarily to direct implantation in electronic materials as required in the microfabrication of devices. With these high-resolution (≤ 0.1-μm) focused ion beams, profiled doping can be achieved in a direct manner, quite unrealizable by any other processes [19a,b]. Also the recent application of mask repair [7] that has been proposed and demonstrated could be of great practical significance in the lithographic process in which masks can still be expected to be used in high-throughput photolithography.

D. Ion Optics and Ion Optical Systems

As pointed out above, the ion optical system that is designed for a given IBL system depends first of all on the source characteristics. In the previous sections the characteristics of the two basic high-brightness sources now available for focused ion probe systems, the GFIS and LMI source, have been discussed and their characteristics are summarized in Table V.

The characteristics of these sources are very suitable for coupling to charged-particle optical systems that focus beams to high-resolution probes. There is a large literature on charged-particle optics for probe and image-forming lenses and systems [75–80], but most of the high-resolution optics have been applied to electron optical systems. The optical systems that have been developed extensively for electron-beam systems use magnetic lenses to achieve high resolution and high current densities in the focused probes. Very advanced EBL systems have also been developed [17,81–83b]. The initial systems started from scanning electron microscope (SEM) systems, and the instruments were modified to incorporate the special requirements for lithography. The EBL systems use probes with lower resolution (larger probe diameter) than the SEM, but they require higher beam currents in the focused probe to allow shorter exposure time and hence faster throughput rates.

While the extensive developments in electron optics could be used quite directly in designing the electron optical system for EBL systems to produce the desired focused probe, the requirement for exposing large areas of the wafer substrates under very exacting dimensional control in lithography also required careful attention to the beam deflection systems. These requirements led to the development of EBL systems with scanning systems that allow large deflections with minimum deflection aberration, very fast blanking, shaped beams, and extremely fine control and very accurate positioning and monitoring of stages that can move rapidly. Much of this development can be applied directly to the development of instruments for IBL with one very important difference. The high-quality magnetic electron lenses used to focus the electron beams cannot be used for focusing ion beams. The refractive power of magnetic lenses is inversely proportional to the mass of the charged particle so the magnetic fields that can be produced with the "round" (axially symmetric) lens and are used in high-resolution electron optics are not strong enough to focus ions of even the lowest mass with adequate resolution. This limitation imposes the use of electrostatic lenses for focusing ion beams. Electrostatic lenses can be used because their refractive power is independent of the mass of the charged particle. In deriving the equations of motion of charged particles in an electric field, the charge-to-mass ratio e/m cancels out, so ions are refracted with the same trajectories as an electron would be when focused with electrostatic lenses. The limit on the refractive power of an electrostatic lens is set by the fields that can be sustained between electrodes rather than the saturation permeability of the magnetic materials used in magnetic lenses. Given this limitation on electrostatic lenses, practical electrostatic lenses have aberration coefficients that are typically an order of magnitude worse than the best magnetic lenses. However, by careful computer-aided design (CAD) the characteristics of electrostatic lenses can be improved for specific applications.

Because of the inherently larger aberrations of electrostatic lenses, ion optical probe systems will have lower resolution than electron-beam systems and consequently lower current densities in the focused probes, assuming electron- and ion-beam sources of equivalent angular current density and energy spread. However, in the total analysis of the instrumental performance many other factors come into consideration. One critical factor is the nature of the ion–substrate interaction that gives two orders of magnitude greater sensitivity of the resists to ions than to electrons. This factor compensates for the higher angular currents of electrons that are now obtained with field-emission sources [52a,b,53,81b].

The optical characteristics of the electrostatic lens system determine the quality of the ion probe that can be realized with the ion sources described above. The chromatic, spherical, and deflection aberrations of the ion optical and scanning system determine the resolution that can be realized in the focused probe, the probe current density, and also the deflection distance using a given ion source and lens system. To first order, a lens will focus an object from a given point to an image at a conjugate point if the refracting medium of the lens is axially symmetric. To focus charged particles, an electric or magnetic field produced by axially symmetric electrodes or pole pieces will satisfy this condition.

The task of designing an electrostatic ion lens is primarily that of choosing a configuration of electrodes and the voltage ratios between the electrodes that produce the desired first-order properties (focal length, position of object and image planes, etc.) with the minimum aberration coefficients. Since the size of the virtual source (see Table V) that is to be focused by the ion optical system is small, the diameter of the image figure obtained in the focused probe will be determined, almost completely, by the aberrations in the ion optical system. Depending on the characteristics of the source and the particular application, either the chromatic aberration, the spherical aberration, or both will set the limit on probe size obtained. Typically the lenses designed to focus the beams from LMI sources with their relatively large energy spread require lenses designed to have minimum chromatic aberration [84,85]. The H_2^+ ion sources should be focused with lenses in which the chromatic and spherical aberrations are balanced [86]. As we have seen, the virtual source sizes have been estimated to be 55 and 5–10 Å for the LMI and GFI sources, respectively (Table V). So if the imaging system were not aberration limited, the optical system would have to magnify the virtual source. However, since the aberration figures expected with the electrostatic lens systems are typically some 10 to 100 nm in diameter, the optical systems are generally designed to demagnify the aberration figure or at least image it at 1 : 1 magnification.

1. Electrostatic Lens Design

We have already noted that the whole subject of charged particle optics is treated extensively in the literature [75–80]. Calculations on a wide variety of lenses, including electrostatic lens configurations, have been reported (see, e.g., Harting and Read [78] and El-Kareh and El-Kareh [76]). Special configurations designed to give lower aberration coefficients [84–88] have been published. These design data can be used to achieve lens designs that would be satisfactory for given instrumental design. However, any advanced design should be developed independently using CAD [80] to determine the characteristics of the optical system for a given instrumental application. The general method will be described briefly.

Starting with a given configuration of electrodes and the potentials on the electrodes, the axial potential $V(z, 0)$ can be obtained by a number of methods; solving the Laplace equation, relaxation net, finite element, or the charge density method are examples of methods that have been used and are described in an extensive literature on the subject [76–80]. The charge density method [78] seems to be the current method of choice. With this method, computer calculations can give $V(z, 0)$ to the desired accuracy, and it is relatively easy to modify potentials or positions of electrodes. If $V(z, 0)$ is known computer calculations give the trajectory of the charged particles from which the first-order optical characteristics and the aberration coefficients can be obtained. Munro [79a] has published a ray-tracing program that is widely used.

A special four-electrode lens was developed using CAD by Szilagyi et al. [87a] and Paik et al. [86] to produce the objective lens in an IBL systems that focused the virtual source to a crossover at the blanking aperture with small chromatic and spherical aberration. They found that, by using four electrodes and operating the lens as an accelerating immersion lens, they could make significant improvements in the optical characteristics of the lens over those of the three-electrode Einzel lenses conventionally used in probe-forming systems. Using the ion optical configuration illustrated in Fig. 13, Paik et al. [86] found that the minimum image figure could be obtained by properly scaling and balancing the magnifications of the objective lens and projector lens, and the two lenses could then be treated as a total lens system. The four-electrode configuration also allows much more flexibility in the operation of the lens; for example, it can be operated in the "zoom" mode, in which the object and image positions are fixed and the accelerating energy of the beam is varied [87a,b]. The four-electrode lens has also been used by Kurihana [88] in a focused ion probe system using an LMI source.

2. Deflection Systems

Since the beam focused by the lenses is deflected in direct write lithography systems, the deflection aberrations must also be calculated. These aberrations are sensitive to the placement and characteristics of the deflectors [89–91]. Electrostatic eight-pole deflector systems on which the electrode potentials have a cosine distribution produce dipole fields that deflect the beam free of second-order aberrations [92]. The placement of these eight-pole deflectors, often referred to as "octopole" deflectors, can strongly affect the deflection aberrations of the projector–deflection system that focuses the beam on the target and scans the probe over it. Postlens deflection is often used, mainly for simplicity of construction, but the deflection aberrations are kept within the needed tolerances only by the long image distance of the projector lens and dynamic correction. In-lens or prelens deflectors can be adjusted to minimize the deflection aberrations. Most of the published configurations designed to minimize deflection aberrations have been developed for electron-beam systems using magnetic lenses with magnetic or electrostatic deflectors [89,90a]. Ohiwa [90a,b] used an in-lens design that can be adjusted to reduce any two of the deflection aberrations to zero or minimize sets of them. Paik and Siegel [91] developed a scheme in which a set of three prelens deflectors can be adjusted to produce deflections with zero aberration coefficients for coma, chromatic deflection, and astigmatism, to third order. The remaining deflection aberrations of curvature of field and distortion can be corrected dynamically by programming the focusing and deflection voltages on the projector lens and deflectors, respectively. This configuration should give a deflection–projector system aberration free to third order that can deflect a focused ion or electron beam over large areas.

E. Focused-Ion-Beam Lithography Systems

1. A High-Resolution IBL System Using an H_2^+ Field-Ion Source

A complete IBL system based on the conversion of an advanced EBL system is under development and construction at Cornell University [93]. The basic system is a prototype of the high-speed, high-resolution EBL system developed by Hewlett-Packard for a production environment [81a,b]. The Cornell group is replacing the electron field-emission source

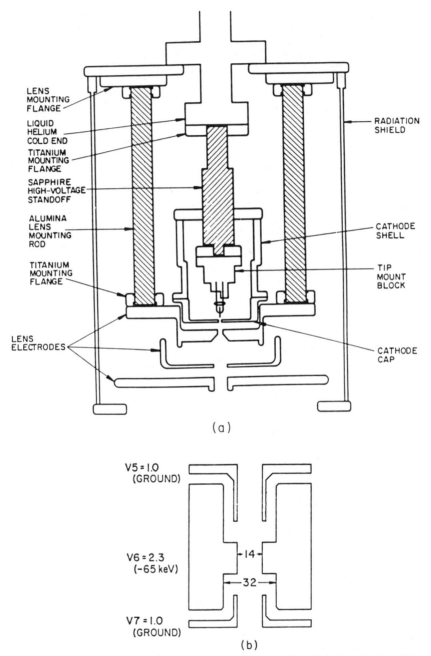

Fig. 23. (a) Schematic of a gaseous field-ion (H_2^+) source with triode focusing lens. [From G. N. Lewis, H. Paik, J. Mioduszewski, and B. M. Siegel, *J. Vac. Sci. Technol. B* **4,** 116 (1986).] (b) Electrode configuration of an Einzel lens projector. [From H. Paik, G. N. Lewis, E. J. Kirkland, and B. M. Siegel, *J. Vac. Sci. Technol., B* **3,** 75 (1985).]

and electron optics with an H_2^+ GFIS based on the source described above (see Section IV.B.4). A specifically designed electrostatic ion optical system of two lenses, alignment deflectors, and blanking and scanning systems makes up the column of the new IBL system. The configuration of the optical system is illustrated in Fig. 13. A schematic of the source and first lens that focuses the probe to a crossover at the blanking aperture is shown in Fig. 23a. A schematic of the Einzel lens projector is given in Fig. 23b. Some modifications of the electronics and the software have been made, although the rest of the system remains essentially intact with a very advanced laser interferometer controlled stage, ultrahigh-vacuum (UHV) system, and advanced mechanical design for vibration isolation and wafer loading and unloading.

The calculated characteristics of this system operating at a final beam energy of 50 keV indicate that with a 2-mrad acceptance half-angle a probe of 10-nm diameter would be produced. Assuming a source with an angular current of 10 μA/sr and an energy spread of 1 eV (FWHM), both of which have been measured for the H_2^+ ion source, the probe would have a current density of ≥ 100 A/cm^2. The postlens deflection system should be able to deflect the beam over 10^4 beam diameters without loss of resolution.

2. Focused Ion Probe Systems Based on LMI Sources

a. Experimental Systems. Several probe systems have been built in various laboratories in the world based on the LMI source. One of the first was built by Seliger and his group at Hughes Research Laboratories [5b,19b]. They later produced a more advanced instrument in collaboration with Levi-Setti at the University of Chicago, who is applying this instrument in a scanning ion mass spectrometer mode for high-resolution analytic studies [94a]. Several systems have been developed and reported by Japanese investigators. Notable are Namba's group at Osaka University [94b], Kumoro [94c] and Kumoro and Kawakatsu [94d], Tsumagari et al. [94e], Miyauchi et al. [94f] and Kato et al. [94g], among others. The Japanese have been very active in developing ion-beam systems based on LMI sources. One system of special interest was reported by Kurihara [88], who used a lens based on the design of Szilagyi et al. [87a] and Paik et al. [86] to produce a probe system that gives a high-resolution probe with substantially higher current density than those obtained with other LMI-based systems. He reported beam currents of $\simeq 25$ A/cm^2 in probes 0.1 μm in diameter and $\simeq 30$ A/cm^2 in a 0.055-μm-diameter probe. By using an advanced four-electrode accelerating design for the first lens, he achieved superior performance compared with that reported for other systems (typically $1-5$ A/cm^2). However, the beam is not focused to a crossover in his

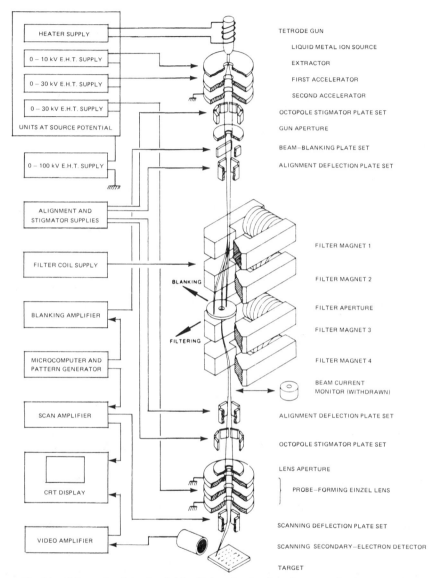

Fig. 24. Schematic of the Cambridge University (Engineering) scanning ion-beam lithography system with a symmetric, magnetic ion filter. [From J. R. A. Cleaver, P. J. Heard, and H. A. Ahmed, *in* "Microcircuit Engineering '83" (H. Ahmed, J. R. A. Cleaver, and G. A. C. Jones, eds.), p. 135. Academic Press, New York, 1983.]

system, and high-speed blanking without spurious deflection effects will undoubtedly be a limitation in this system design. Cleaver *et al.* [94h] produced a prototype system at Cambridge University that uses a magnetic sector filter to separate the ion species from an LMI source when used with an alloy. This system is illustrated in Fig. 24.

b. Commercial Ion Probe Systems. A number of companies are now producing focused ion probe systems based on LMI sources. These instruments have a variety of ion optical systems. The features that all have in common are a mass filter to separate the various ion species from alloy sources and a laser interferometer controlled stage to provide the means of carrying out the processes involved in the microfabrication of electronic devices.

Figure 25 is a schematic of the IBT MICROFOCUS system produced by Ion Beam Technologies, Inc., Beverly, Massachusetts. They use a relatively low voltage beam through the EXB mass analyzer and then accelerate the beam to 150 keV as it is focused on the target. The figure also illustrates in block diagram some of the electronic components required to drive the total system. A schematic of the ion optical column of the VG Semicon (E. Grinstead, England) focused ion beam system, IBL-100S, has already been shown in Fig. 14. Other commerical ion probe systems are manufactured by Japanese Electro Optical Laboratories (JEOL) and MBI Micro Beam, Inc. (MicroBeam). The latter instrument is designed to be a scanning ion microscope and analytical tool.

F. Parallel Processing with Ions

Two methods have been used to pattern resists in a parallel mode: proximity masks and ion projection systems. Sources such as the duoplasmatron [18] are used in these methods and provide the potential for very high throughput lithography given the high sensitivity of resists to ions. Ion beam processing is used in other steps in the microfabrication of electronic devices (e.g., ion etching and milling, ion implantation of dopants), usually through patterned resist layers that act as stencil masks. These procedures are regular steps in current microfabrication processes and are used extensively and widely reported in the literature [18]. We shall limit our discussion to the lithographic methods that have been reported for the parallel patterning of resists by ions, either through masks or by a projection system.

1. Masked Ion-Beam Lithography

The group at Hughes Research Laboratory have developed methods for parallel patterning of resists using proximity printing through specially

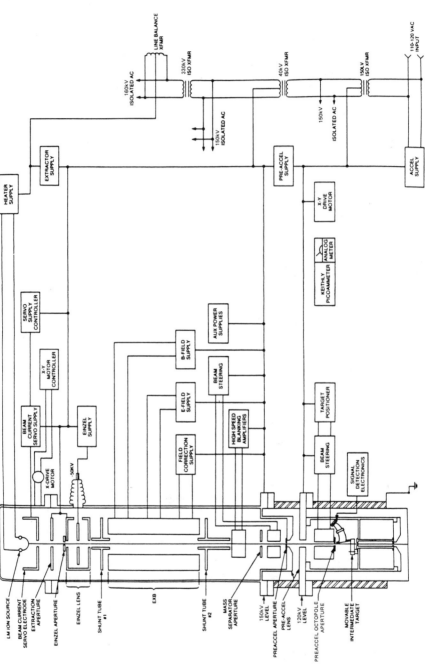

Fig. 25. Schematic of the Ion Beam Technologies, Inc. IBT MICROFOCUS system showing the column and block diagram of the control electronics. Courtesy Ion Beam Technologies, Inc., Beverly, Massachusetts.

fabricated masks [19b,45]. A collimated beam of protons generated in a duoplasmatron source with energies ranging from 130 to 175 keV were used to expose resist-covered silicon wafers. They investigated the characteristics of the process and determined the conditions necessary for obtaining submicrometer (0.1-μm) resolution. Perhaps the most critical element in this development is the mask technology. The mask must be rigid enough to be self-supporting and not be distorted beyond specifications under the thermal load of the ions absorbed in the masked areas. To achieve the desired goal of 0.1-μm edge resolution it was necessary to determine the angular scattering of the ion beam at the mask to obtain the maximum separation that could be tolerated between the mask and substrate. They determined [45] that a thin membrane (≈ 0.7 μm) of $\langle 100 \rangle$ silicon properly oriented to produce channeling of the H$^+$ beam along the axis caused only a small angle ($\approx 0.4°$) scattering of the proton beam. Thus, a mask held to a separation of less than 14 μm from the substrate would maintain the 0.1-μm edge resolution. To obtain the necessary rigidity in the membrane the silicon was heavily doped with boron, which produces tensile stress in the membrane. This doping has the additional feature of limiting the etching process involved in the preparation of the thin membrane. The absorbing pattern of the mask was a 0.7-μm layer of gold (with a 500-Å chromium adhesion layer). The gold pattern was produced by electron-beam resist exposure, reactive ion milling, and sputter deposition. The characteristics of this mask were investigated both experimentally and theoretically [19c]. The structural effects of mask heating can result in both out-of-plane membrane deflection and in-pattern distortions. The modeling calculations of the Hughes group indicated that the boron-doped silicon membranes should have a tensile stress of greater than $1-2 \times 10^9$ dyn/cm^2 to avoid out-of-plane deflections. The size of mask that can be tolerated is a function of the heating of the mask by ions absorbed in the gold and the part of the beam that is absorbed in the channeling through the open areas. The heating is, of course, a function of the incident beam flux and the fraction of the mask covered by the gold pattern. The Hughes group found that, at a flux of 0.1 W/cm^2 (~ 1 μA/cm^2), a silicon membrane with 50% coverage of gold could be as large as 1.4 cm^2 without suffering in-plane distortions of greater than 0.1 μm.

Bartelt et al. [95] also demonstrated the feasibility of this masked ion-beam lithography (MIBL) process by fabricating NMOS test chips using their masks for each of the four levels (thin oxide, polysilicon gate, contact windows, and metallization). The test chips contained submicrometer gate transistors with gate width-to-length ratios varying from 0.27 to 23. The current–voltage characteristics of these MIBL-fabricated transistors were found to be comparable to similar transistors fabricated by EBL. To evaluate possible effects of radiation damage from the ion exposures, $p-n$ junctions and capacitors were incorporated into the test chip. Only low leakage was observed (< 1 μA), and reverse breakdown was at a voltage of $\simeq 12$ V. Threshold voltage shifts of less than 0.1 V and low surface state densities were indicated by $C-V$ measurements on the thin oxide capacitors before and after bias-temperature stress. Thus, the transistors, diodes, and capacitors show little or no radiation damage from the ion-beam exposures. These important observations provide another demonstration of the feasibility of fabricating actual submicrometer electronic devices by IBL.

The group at Lincoln Laboratory, MIT, have also been investigating the feasibility of MIBL using specially designed masks. They first fabricated masks made only with polyimide, absorbing the H^+ ions in the thicker regions of the polyimide and etching the transmission areas thin enough to transmit the 50- to 200-keV protons [96a]. They achieved periodic patterns with lines ~ 600 Å wide spaced 0.3 μm apart. This group also developed stencil masks made of silicon nitride and produced 80-nm lines in grating patterns with 320-nm periods [96b]. More recently [96c] they have developed a grid support mask that can provide a stable mask with arbitrary pattern definition.

2. Ion Projection Lithography

An ion-beam system designed to project an image of a stencil mask on a substrate at a reduced magnification has been under development by a group in Vienna, Austria [97a,b,98]. The latest prototype system (IPLM-01) reported by Ion Microfabrication Systems (IMS) by Stengl *et al.* [98a,b] indicates significant progress, and a variety of test demonstrations have shown the great potential of ion projection lithography.

A highly simplified schematic of the IMS instrument is illustrated in Fig. 26. IMS uses a duoplasmatron to produce beams of several ion species, H^+, He^+, N^+, Ar^+, or Xe^+, with a mass analyzer to obtain the desired ion species. The beam from this source is accelerated to only $5-10$ keV before illuminating the stencil mask. The immersion lens accelerates the ions to 60 to 90 keV, and a projector lens produces a demagnified image (1 : 10 or 1 : 5) of the mask on the substrate. The octopole deflector system and solenoid system between the immersion lens and projector are used to align and rotate the image to obtain accurate registration of the patterns on the substrate. The ion optical system has a very small angular aperture, and therefore the depth of focus is greater than 100 μm, avoiding problems caused by distortion in the substrate.

The prototype system takes a 25-mm-diameter mask, so the image field

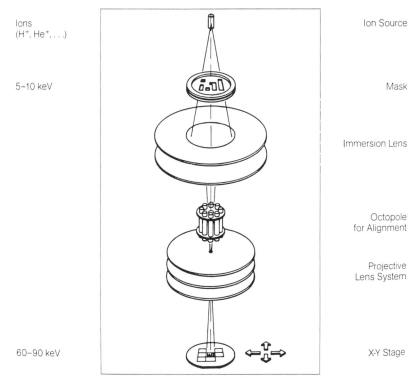

Ions
(H⁺, He⁺, ...)   Ion Source

5–10 keV   Mask

                                                                     Immersion Lens

                                                                     Octopole
                                                                     for Alignment

                                                                     Projective
                                                                     Lens System

60–90 keV   X-Y Stage

Fig. 26. Block schematic of the ion projection lithography machine by Ion Microfabrication Systems, Vienna, Austria. Model IPLM-01. [From G. Stengl, H. Loschner, W. Maurer, and P. Wolf, *J. Vac. Sci. Technol. B* **4,** 194 (1986).]

is 2.5 mm in diameter. A step-and-repeat procedure is used to expose the total wafer area. The mask is irradiated with a relatively low ion current density of 0.1 to 10 μA/cm² with 5-keV energy. The power density on the mask is therefore some 1000 times lower on the mask than at the substrate, minimizing heating, sputtering, and irradiation damage. Stengl *et al.* [98a,b] found that a 2-μm-thick nickel mask could be continuously irradiated for some 250 hr. Since the mask is imaged at 1 : 10 demagnification, there is a 100× gain in current density on the substrate, producing dose rates as high as 1 mA/cm² sec, or almost 10^{16} ions/cm² sec. Figure 27 illustrates the chip exposure times required for different processes. Lithography using a PMMA resist (sensitivity 1 μC/cm²) would require only 0.01–0.1 sec per chip area, and a 4-in. wafer could be exposed in some 1 to 2 sec plus the overhead time required for stepping. This very high throughput rate makes ion projection lithography a very promising method for future microfabrication production. The range of exposures required for a

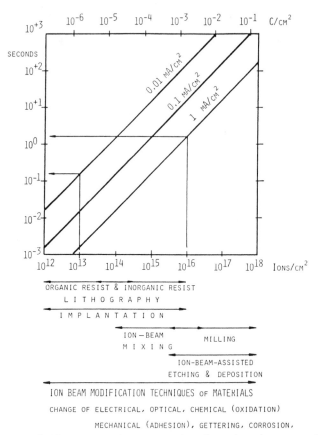

Fig. 27. Plots of chip exposure times versus dose for various processes using the ion projection lithography machine (magnification, reduction 10:1, chip size). [From G. Stengl, H. Loschner, W. Maurer, and P. Wolf, *J. Vac. Sci. Technol. B* **4**, 194 (1986).]

wide range of ion-processing procedures such as implantation, milling, ion-beam-assisted etching, and ion modification of materials is shown in the lower part of Fig. 27.

The IPLM-01 is in the development stage. While 0.2-μm resolution has been demonstrated, several areas must still be evaluated and developed further. The limitations introduced by the mechanical shutter used to blank the image, by the mask, and by image distortion still have to be evaluated. Final critical resolution and registration tests have not yet been reported. However, for mass production of submicrometer devices ion projection lithography rivals and may well be superior to x-ray lithography.

V. PROBLEMS AND LIMITATIONS IN ION-BEAM LITHOGRAPHY

While focused or maskless IBL is analogous to EBL, the heavier masses of the ions introduce differences that are both advantageous and problematic. The problems that can set limitations on IBL are related to (1) the relatively slower velocity of the ions and (2) the fact that their kinetic energy is deposited in the target in a much smaller volume. The first characteristic requires special treatment of the blanking and scanning systems. The slower velocities also introduce the possibility of greater Coulombic interactions and hence energy broadening and transverse spreading in the focused ion beam. The second characteristic means that organic resists such as PMMA are so sensitive to ions that they are fully exposed by very few ions, especially high-resolution pixels (100–200 Å diameter). The statistical fluctuation from pixel to pixel can be very large, and therefore the "shot noise" is large. Each of these effects must be considered, and the methods for overcoming the limitations they introduce must be evaluated.

A. Blanking of Ion Beams

Brown *et al.* [99] suggested that in the application of focused IBL an unacceptably slow blanking rate is imposed by the relatively slow velocity of ions. However, this limitation can be overcome by the proper design of the blanking system. The blanking rate that can be realized then depends only on the mechanical and electronic tolerances achieved in building and driving the blanking system.

When the period of the blanking signal becomes comparable to the time of flight of the charged particles through the blanking plates, two problems occur. The beam deflection across the knife edge used to block the beam decreases, and the beam is spuriously deflected on the target, causing unwanted exposure. Both of these problems can be eliminated by the use of a slow-traveling waveguide deflection electrode [75] that tracks the particle as it passes through the blanker. Although there is no intrinsic limit on the blanking rate with this method, such a traveling waveguide would require a very complex system. A simpler, more practical solution is to use sequential pairs of parallel deflector plates with suitable time delays between the application of the blanking signal to each pair of plates. Kuo *et al.* [100] described such a system for an EBL system.

Paik *et al.* [46] evaluated and adapted this scheme for blanking in focused-ion-beam systems. A schematic of their proposed system using two pairs of deflector plates is shown in Fig. 28. The blanking system is located

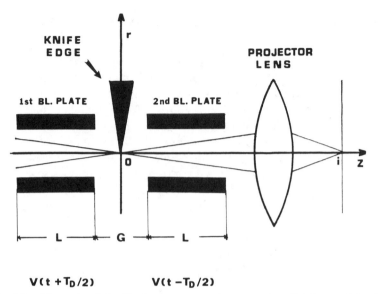

Fig. 28. Schematic of the blanker geometry using two pairs of deflector plates. The crossover focused by the first lens is at the knife edge, and the projector lens focuses this crossover on the substrate. [From H. Paik, Ph.D. Thesis, p. 120. Cornell Univ., Ithaca, New York, 1985.]

so the knife edge is at the image plane of the first lens, and this image is also at the object plane of the projector lens so that deflected charged particles appear to be deflected about a center at the knife edge. Here, V is the blanking voltage and T_D the delay time between the two pair of plates of length L with a gap G. Assuming a sine wave blanking voltage, the deflection across the knife edge will be proportional to $\mathrm{sinc}(\pi f T_L)$, where f is the blanking rate and T_L the time of flight of the charged particle through the plate of length L. To prevent the blanking from being pattern (and hence frequency) dependent, the deflection across the knife edge should not vary with frequency up to the maximum allowable frequency. If up to a 10% decrease in deflection is allowed,

$$f_{\max} < 1/4 T_L.$$

Therefore, it is desirable to keep the plate length as short as possible. Shortening the plate decreases the deflection sensitivity, which can be compensated for by increasing the gap G between the plates. The total deflection across the knife edge is

$$\Delta x_{KE} = V_D (L^2 + LG)/4 V_0 D,$$

where V_0 is the voltage to which the beam is accelerated and D the distance and V_D the voltage between the deflector plates.

A spurious deflection on the target occurs during the rising and falling edge of the blanking voltage, limiting the resolution of the probe system. This effect can be calculated numerically [46] by tracing a ray through the blank and extrapolating back to the knife edge, which is at the object plane of the projector. The axial displacement in this plane multiplied by the magnification of the projector is the spurious deflection on the target–image plane. Paik [101,102] has shown that this spurious deflection can be determined analytically to a very good approximation. The spurious deflection is found to be directly proportional to the static deflection on the knife edge, which therefore should be kept as small as possible. Assuming a deflection limited to 1 μm at the knife edge for beams with a maximum beam diameter of 0.1 μm and designing a blanking system with dimensions approaching the limits of mechanical tolerances that can be achieved with care, the calculated resolutions as a function of blanking rate are shown in Fig. 29. Plots are shown for focused probes of electrons, H_2^+, and Ga^+, each with the delay time T_D adjusted to within accuracies of ± 0.5 and $\pm 0.1\%$. The probe size (resolution) is taken to be 10 times the spurious

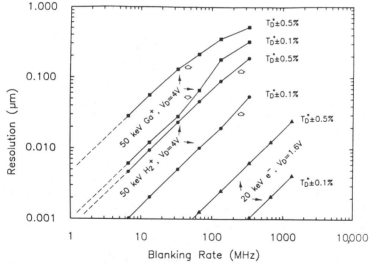

Fig. 29.   Resolution (beam diameter) as a function of blanking rate for 50-keV Ga^+ ions, 50-keV H_2^+ ions, and 20-keV electrons using two pairs of deflector plates of length 0.2 cm, gap 0.8 cm, and separation 0.04 cm. The deflection across the knife edge is taken as 1 μm, and a projector magnification of unity is assumed. Two plots are shown for each of the charged particles with different errors in the time delays T_D^* between the voltages V_D on each set of plates. [From H. Paik, Ph.D. Thesis, p. 138. Cornell Univ., Ithaca, New York, 1985.]

deflection. In practice the extreme sensitivity to the delay time may require adjusting this parameter by varying the beam energy and hence the time of flight of the charged particles through the blanking system. The optimum delay times in this system are 0.1243, 4.6315, and 27.10 nsec for electrons, H_2^+, and Ga^+, respectively, taking the velocities of electrons, H_2^+, Ga^+ as 0.1227 nsec/cm 20 keV), 4.570 nsec, and 26.73 nsec/cm (50 keV), respectively.

As Fig. 29 illustrates, the blanking rates of 100 to 300 MH_3 can be achieved with the proposed system for H_2^+ ions depending on the resolution required. This rate is not as high as that which can be achieved with electrons but considerably better than can be achieved with heavier ions such as Ga^+. Gallium ions produce another problem, because gallium has two isotopes with atomic weights of 68.925 and 70.924 amu, which would require different delay times and therefore introduce an error that could not be corrected. In Fig. 29 isotopically pure gallium (68.925 amu) was assumed.

B. "Shot Noise": Limitations Introduced by Statistical Fluctuations

The much greater sensitivity of resists to ions than to electrons was discussed earlier. For example, PMMA is 50–100 times more sensitive to 50-keV protons than to 20-keV electrons. (PMMA would be less sensitive to 50-keV electrons.) Consequently, PMMA is fully exposed by doses of H^+ ions of 0.5 to 1 $\mu C/cm^2$. If a high-resolution focused H_2^+ ion probe such as that described in Section IV.E.1 were used to expose PMMA, only a very few ions would fully expose each pixel. A pixel 200 Å in diameter would be exposed by only ~ 10 H_2^+ ions. (Every H_2^+ ion generates two H^+ ions, each with half the energy of the incident H_2^+ ion [49b].) The focused H_2^+ probe could have current densities of > 100 A/cm^2, and thus an exposure of ≃ 5 nsec would be required. However, the statistical fluctuation is given approximately by the square root of the number of primary particles, so in sampling with so few particles the variation between pixels could be large (i.e., 10 ± 3). The situation would be even "noiser" at 100-Å resolution, the resolution we hope to achieve with ion beams; here, some 2.5 ± 1.6 H_2^+ ions would be adequate to expose these small pixels. At larger dimensions, for example, 0.1-μm-diameter pixels or lines in PMMA, the number of H_2^+ ions required would be only ≃ 250 ± 16, a fluctuation of more than 6%, which would not be acceptable for EBL but, as we shall see, could be for IBL.

The situation, however, is not as hopeless as it may seem from these large statistical fluctuations. There is an extremely wide latitude in allowed dosage with light ions like H_2^+. All the energy deposited by each incident ion is confined to a very narrow spatial range, and so it may be possible to overdose or make multiple exposures until there would be a very low probability of any pixel not receiving an adequate dose even though it received the largest deviation in exposure from the mean. There is a limit on allowed dose, for the PMMA could start to act as a negative resist at some level. These exposure characteristics of resists to ions must be determined experimentally. The modeling experiments discussed in Section II.D.1 and the developed contours illustrated in Figs. 8 and 9 indicate that, because of the narrow confinement of the energy deposited, the resist can also be greatly overdeveloped without causing the lines to broaden.

When high-resolution focused H_2^+ ion beams are available, it will be important to determine experimentally what latitudes are possible in the exposure and development of resists. The contours illustrated in Figs. 8 and 9 postulated unrealistic square beams. Actual beams focused by systems such as that described in Section IV.E.1 produce aberrated images that have tails. These tails will expose the resists with a much lower dose that is spread out from the central exposed line. Paik [102] studied this effect for the case in which a probe is focused by this ion optical system using a acceptance half-angle of 8 mrad. The beam would be mainly spherically aberration limited with some broadening caused by chromatic aberration and have a diameter of 500 Å (FWHM). Defocusing the beam so the substrate is at $z_i = -0.75 C_{si} \alpha_i$, where the normalized axial position $z_i = 0$ is the Gaussian plane and C_{si} and α_i are the spherical aberration coefficient and angular aperture in image space, respectively, produces an image figure with minimum tails. The current density in the beam at the substrate is shown in Fig. 30a for various acceptance half-angles α_0 at the defocus $z_i = -0.75 C_{si} \alpha_i$. The energy deposited in the resist obtained by convoluting the beam for $\alpha_0 = 8$ mrad with the ion interaction with the resist is shown in Fig. 30b. As is to be expected, carrying through the full CONDEV modeling program gives contours that spread with extended development times (see Fig. 30c). At the longest development times, even the ends of the tails have produced enough exposure to cause the resists to develop. There is a definite limit to the width for any given exposure, however, and by controlling the development time one can obtain narrower lines. Of course, by using the smaller acceptance angles (e.g., $\alpha_0 = 2$ mrad), one would avoid the effects of the broad tails.

Another approach to this problem of the statistical fluctuations inherent in exposing sensitive resists to ions is to use less sensitive resists. Resist sensitivity may no longer be a primary concern in the development of new resists for IBL. By concentrating on the development of resists with properties that are superior for the subsequent transfer and handling processes required in the fabrication of complex submicrometer electronic devices, it

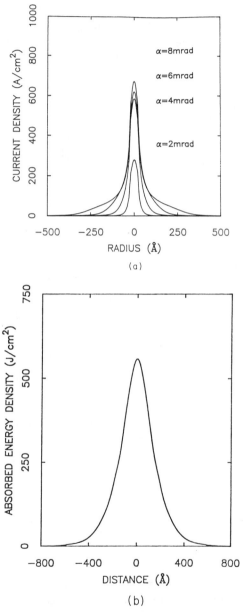

Fig. 30. (a) Current distribution in a probe focused by the ion optical system described in Section IV.E.1 for different acceptance half-angles. The current distribution has been calculated for the image defocused to $z = -0.75 C_{si}\alpha_i^2$, where the Gaussian image plane is at $z = 0$; C_{si} and α_i are the image spherical aberration coefficient and image convergence half-angle, respectively. (b) Absorbed energy distribution in PMMA at a depth of 2000 Å calculated by

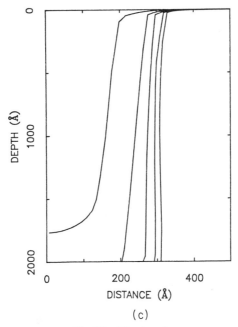

(c)

Fig. 30 *(Continued)*

convoluting the current distribution for $\alpha = 8$ mrad shown in (a) with the energy loss profile in PMMA for 25-keV H^+ at a depth of 2000 Å calculated by fitting the results obtained with the PIBER program. H_2^+ ions at 50-keV energy act on entering a solid (resist) as if they were two independent H^+ ions, each with one-half the energy of the incident H_2^+ ion. (c) CONDEV program used to calculate development contours obtained using absorbed energy in PMMA illustrated in (b) after development times of 1, 3, 5, 7 and 9 min in saturated 1:1 methyl isobutyl ketone/isopropyl alcohol (MIBK–IPA). The peak dose was taken to be 2.5×10^{-6} C/cm^2 of H^+ ions. [From H. Paik, Ph.D. Thesis, Cornell Univ., Ithaca, New York, 1985.]

would be possible to realize a positive gain. Research and development of new resists that could optimize the properties needed in pattern transfer and the gamut of the processing steps required in microfabrication without concern for the sensitivity factor may be one of the most important aspects of IBl as we move into fabricating structures with dimensions in the low submicrometer and nanometer ranges.

C. Coulombic Interactions

The charged particles in beams repel one another, and this Coulombic interaction can play an important role in limiting the resolution and current density to which an electron or ion beam can be focused. Boersch

[103] first described an energy broadening of electron beams at high current densities. Early attempts to explain the "Boersch effect" [103,104] were based on the assumption that there is an energy exchange originating from collective interactions with a large number of charged particles that produces space charge oscillation. While such collective interactions undoubtedly occur in the case of very high current density beams and in magnetically confined long parallel beams, in low-current-density beams with only short regions of high current density (e.g., at focus) the number of charged particles is smaller and the problem can be treated as collision between individual particles. This condition obtains for the ion beams we consider in the focused-ion-beam lithography systems in this chapter. Gomer [69] postulated "space charge" as the effect producing the energy broadening in LMI sources. However, his theory has met with considerable challenge [66a,b,70].

Rose and Spehr [105] published an extensive review of this subject describing both the analytical methods and Monte Carlo computational treatments that have been applied to the Boersch effect. While many investigators limit the term "Boersch effect" to the energy broadening in particle beams first described by Boersch, Rose and Spehr also include the beam broadening that occurs under the Coulombic repulsion in charged-particle beams, naming it the "spatial Boersch effect." Both effects are caused by the same basic physical phenomenon: the collision and transfer of momentum between charged particles in a beam. The interactions that change the transverse velocity of the particles with respect to the axis cause a broadening of the beam and the size and shape of the focused image. Interaction parallel to the optical axis of the beam results in longitudinal spreading of the charged particles, causing an energy broadening in the beam. This energy broadening produces increased chromatic aberration in the image focused by the lens system. Considerable attention has been given to developing analytical methods for determining the energy-broadening effects [106–109]. Rose and Spehr also give a detailed report of their analytical method in their 1983 review article [105].

The analytical treatments by Rose and Spehr [105,106] apply to both electrons and ions, but the examples they use are related primarily to the conditions that would obtain in an electron-beam probe system in which the source is demagnified. They evaluate the effect of a beam crossover and the Coulombic repulsion that occur for given distances before and after the crossover. In this situation the mean axial distance between particles is large compared with the impact parameter, and if the crossover diameter is less than 10 times the impact parameter, the crossover can be treated as a point and there are no multiple scattering effects over the whole region.

The mean quadratic energy broadening under these conditions is proportional to the beam current and to the square root of the mass of the particle. Therefore, a beam of H_2^+ ions would have the same mean quadratic energy broadening as a beam of electrons with a current some 60 times that of the ions. Since resists are some 100 times more sensitive to H_2^+ ions than to electrons, energy broadening in ion beams should not be adversely affected if the focusing conditions are the same.

However, the ion optical systems that are being designed for IBL systems generally magnify the virtual source and have smaller angular convergence angles than the electron-beam systems. The limiting conditions used by Rose and Spehr no longer obtain as the angular convergence angle $\alpha_0 \rightarrow 0$. When there are very small convergence angles over a long distance in an optical system focusing charged particles, there can be strong Coulombic repulsions that produce spatial spreading of the image. While Rose and Spehr present a short analytical treatment of this spatial Boersch effect in an appendix, Monte Carlo calculations have provided us with the most useful data on this effect so far.

The energy-broadening Boersch effect can also be investigated by Monte Carlo calculations, and most treatments evaluate both the spatial and energy effect caused by the Coulombic repulsive forces between the charged particles [110,111a,b]. The work that is most pertinent to our considerations for focused-ion-beam systems has been reported by Groves *et al.* [112] and Yau *et al.* [113]. While energy broadening of charged-particle beams can also be calculated for special geometries to a sufficient degree of accuracy by Monte Carlo computer simulations, the calculations require a large amount of computer time, and the empirical relations obtained do not give complete information as the functional dependencies or elucidate the underlying physical processes.

Groves *et al.* [112] treat the case of the broadening of the electron image of a field-emission source produced by a single lens. They start with a random initialization of the electrons emitted from the source and obtain random vector velocities within the beam cone with a chosen energy spread. These random electrons are followed as they are accelerated to their full energy with all interaction "turned off." Thus, the electrons appear to come from a point source with full beam voltage and with distributed positions and velocities that are determined primarily by the beam current. The electrons are traced from the source through the lens and to the image plane using system parameters that would be typical for obtaining a simple focused electron-beam probe (beam voltage 20 kV, column length 30 cm, magnification unity). Their results are given as plots of spot diameter as a function of beam current, semiangle subtended by the lens aperture, and column length. The focused spot diameter was found to vary roughly linearly with beam current and column length and inversely with the 1.25 power of the aperture semiangle. The effect of chromatic aberration on the image figure was also determined assuming a simple thin magnetic lens. The interesting and surprising result was that for this electron beam the spot broadening for low electron beam currents (<5 nA) was almost totally caused by energy broadening in the beam with the consequent chromatic aberration. At higher currents (>10 nA), energy broadening no longer produced the dominant effect, and transverse or spatial effects determined the diameter of the focused spot. These observations obtained with Monte Carlo simulations were confirmed experimentally by Groves *et al.* [112].

The same Monte Carlo method was applied to focused ion beams by Yau *et al.* [113]. Their method was applied both to a Ga$^+$ ion beam from a typical LMI source and to H$_2^+$ and H$^+$ from a field ionization source. However, now with ion beams they were concerned with the possible energy-broadening effects that might occur at the source in the acceleration space because in that region the charge density could be highest and the particles initially move slowest. Their computations indicate that a Ga$^+$ source with a 350-mrad total emission semiangle and an accelerating field of 20.5 V/Å has an energy spread of 5.0 eV. For the H$_2^+$ and H$^+$ source they took a 20-mrad emission halfangle and a field of 1.0 V/Å and calculated an energy broadening of 4.0 eV. However, Hanson and Siegel [50] measured the energy width of the H$_2^+$ ion beam from their source and found it to be 1.0 eV (FWHM). This discrepancy throws serious doubt on the method or on the initial conditions used byYau *et al.* [113] in calculating their source characteristics. The error could very likely have been introduced by the high total emission current they chose (100 times the current transmitted by the aperture).

The effects they calculated for the focusing of the ion beams by a single lens should still be valid, for they used the same method as that used by Groves *et al.* [112] substituting the appropriate values of q/m. One of the approximations they made that neglects effects to which the electron beam was not subjected is the variation of the potential as the ions passed through the electrostatic lens. They assumed a thin weak Einzel lens that ignores the very large changes in the potential that exist in real electrostatic lenses. However, this approximation should not produce a significant error in the results obtained if accelerating lenses are used as they are in the IBL system described in Section IV.E.1. Even in decelerating lenses the ion trajectories are rather far apart over the relatively short lengths they are being refracted. The results of Yau *et al.* [113] are shown in Fig. 31. The

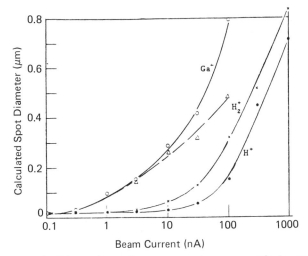

Fig. 31. Curves of minimum beam diameter versus beam current for beams of Ga^+, H_2^+, and H^+ using a single lens focusing the beams from their respective sources with a 2-mrad half-acceptance angle. The dashed line shows the contribution of the chromatic aberration to the diameter of the gallium probe, clearly the dominant effect up to a 10-nA beam current. [From Y. W. Yau, T. R. Groves, and R. F. W. Pease, *J. Vac. Sci. Technol., B* **1**, 1141 (1983).]

energy broadening in the Ga^+ beam produces considerable chromatic aberration in that image, as shown by the dashed line, limiting the resolution that can be realized by focusing the beam from such an LMI source. Experimental results confirm this observation for the IBL systems based on LMI sources. All the systems that have been described typically have focused probe diameters of some 0.1 μm. Smaller beam diameters are obtained only at significant reduction of the beam currents. On the other hand, according to Yau *et al.* [113] the H_2^+ or H^+ beams have computed spot diameters that do not increase until the beam currents exceed 5 and 10 nA, respectively. These beam currents would be quite adequate for IBL. For example, the resists such as PMMA used for electron-beam exposures are $\simeq 100$ times more sensitive to hydrogen ions than to electrons and a 5-nA beam current of H_2^+ ions would be equivalent to 1-μA beam current of electrons. For probes focused to nanometer dimensions (e.g., a 50-nm-diameter spot) an H_2^+ ion current of 5 nA giving two charges per ion in the resists would have a dose equivalent of some 500 C/sec cm^2 requiring a 2-nsec exposure per pixel for PMMA, which has a sensitivity of 1 μC/cm^2. A 1-nA beam current of H_2^+ ions focused to a 10-nm-diameter probe would give a dose of $\simeq 2500$ C/sec cm^2. Since the very low number of ions needed to expose such a small pixel if the sensitivity of the resist were of the order of 1 $\mu C/cm^2$ would cause severe statistical fluctuations, a resist of much lower sensitivity is indicated. These H_2^+ ion-beam currents should be below the levels where significant Boersch effects are to be expected.

VI. CONCLUSIONS

If the dimensions of VLSI systems and microelectronic devices are to be extended to have elements in the nanometer range (<0.1 μm), IBL is a method with great potential. The characteristics of light-ion exposure of resist is such that a focused beam of H_2^+ or H^+ ions could be used to structure patterns of 10-nm resolution on thick substrates with very close spacings since there are no proximity effects. What is needed at this stage is the completion of an experimental very high resolution IBL system that can demonstrate this potential and determine experimentally the ultimate characteristics of IBL. New resists could be developed that have superior resolution and processing characteristics without concern for high sensitivity.

The fabrication of structures with very small dimensions is at the forefront of current research [1,114]. While there is skepticism about the feasibility of miniaturizing conventional electronic devices into the nanometer range because of the effects of interconnects and cross-talk, the capabilities of focused ion beams to produce well-defined profiles of implantation and detailed structuring could offer new approaches to these problems. An area of development using nanometer structure is the fabrication of very high speed GaAs MESFETs with short gates [115]. The most exciting potential lies in the possibilities of new devices based on the quantization effects that occur when dimensions approach 10 nm. Mankiewich et al. [116] at AT&T Bell have made nanometer MOSFETs with very narrow inversion layers to study quantum conduction. The group at Yale is also fabricating nanometer structures to study quantization effects and have demonstrated one-dimensional electron localization in narrow aluminum wires [117]. At MIT, gratings have been fabricated into MOSFET device structures to study the effects of periodic potential modulation and of quantum effects from two-dimensional confinement of the electronic density of states. They hope that these studies could lead to the development of nanometer-dimensional quantum-based devices and systems that could provide higher density and functionality than transistor-based systems [118].

These nanometer structures, however, were fabricated by EBL. The instrumentation and technology for EBL have been developed in several laboratories to a level where 10- to 20-nm lines spaced about 40–70 nm apart can be produced [115–117]. While most of the experimental development of electron-beam structuring at nanometer dimensions has been carried out on modified SEMs and STEMs, some manufacturers such as JOEL, VG Scientific, and Cambridge Instruments are marketing EBL systems designed to pattern in the nanometer range.

These systems are extremely slow and are only being used for studies of small experimental devices requiring very limited amounts of patterning at nanometer dimensions. Throughput is not yet a consideration and EBL with the limitations set by the very low sensitivity of resists to high-voltage electrons is not a serious constraint. Mackie and Beaumont [115] report that to obtain 10 nm continuous lines using PMMA as a resist, a dose of 6330 $\mu C/cm^2$ was required and the minimum pitch they could obtain was 40 nm in writing these grating structures. When devices having some structures of nanometer dimensions come to be fabricated on a large scale or even for a small number of custom devices, throughput will be an important factor.

Smith [119] has analyzed the pixel transfer rate for various lithographies based on the statistical noise limitation inherent in the exposure statistics for given minimum linewidths and linewidth control. The rates he obtained for writing 0.5-μm-wide lines with ion and electron beams were of the same order of magnitude. However, we have calculated, using the Smith analysis, that when writing 50-nm lines, IBL would have a limiting rate two orders of magnitude faster than EBL (10^8 Hz versus 10^6 Hz). The greatly increased data rate available with IBL will make this the required method when even small numbers of custom devices are to be fabricated.

For large numbers of devices parallel processes will be required and parallel processing systems, whether photolithography, x-ray lithography, ion projection lithography, or proximity MILB, require masks. Focused-ion-beam lithography will be the fastest method, perhaps the only satisfactory method, for producing masks with nanometer structures. Primary or "mother" masks for complicated VLSI structures could be produced in reasonable exposure times, much faster than EBL patterning, if a significant fraction of the pattern requires nanometer dimensions. Daughter masks could be fabricated by parallel methods. The parallel method of choice for high throughput could well be ion projection lithography using the approach of Stengl et al. [98a,b]. Here the mask is imaged at a 10:1 reduction so the low ion energy load on the mask allows exposure of resists at dose rates as high as 1 mA/cm^2, and the time to expose a full wafer is set by the step-and-repeat rate at which the stage can be moved. Smith's analysis [119] concluded that deep UV would be the fastest for structures down to $\simeq 0.5\ \mu$m, and below that the synchrotron radiation source would take over. However, he considered only proximity mask methods for IBL where the dose rate is limited by the mask tolerance to heating. As we have pointed out (Section IV.F.2) the ion projection system allows a $10^2 - 10^3$ greater exposure rate and the statistics have to be reevaluated. Our calculations indicate that ion projection lithography would provide the greatest transfer rates, exceeding those available with synchrotron radiation.

It is well to remember that focused-ion-beam systems have very important applications other than patterning in resists for IBL. The H^{2+} system could be used in material modification at high resolution in device manufacture. The applications of the instruments based on LMI sources producing a wide variety of ion species that can be used, for example, for implantation and modulating doping of device structures, mask repair, and ion-beam-assisted reactive ion etching. These applications are so attractive that there are a number of commercial instruments available (see Section IV.E.2.b).

REFERENCES

1. R. E. Howard and D. E. Prober, *in* "VLSI Electronics: Microstructure Science" (N. G. Einspruch, ed.), Vol. 5, p. 145. Academic Press, New York, 1982.
2. B. M. Siegel, G. R. Hanson, M. Szilagyi, D. R. Thomas, R. J. Blackwell, and H. Paik, *in* "Submicron Lithography" (P. D. Blais, ed.), Vol. 333, p. 152. Soc. Photo-Opt. Instrum. Eng., Bellingham, Washington, 1982.
3. R. Clampitt, K. L. Bitten, and D. K. Jefferes, *J. Vac. Sci. Technol.* **12,** 1208 (1975).
4. R. Clampitt and D. K. Jefferes, *Nucl. Instrum. Methods* **149,** 734 (1978).
5a. R. L. Seliger, J. S. Ward, V. Wang, and R. L. Kubena, *J. Appl. Phys.* **34,** 310 (1979).
5b. V. Wang, J. W. Ward, and R. L. Seliger, *J. Vac. Sci. Technol.* **19,** 1158 (1981).
6. R. L. Kubena, C. L. Anderson, R. L. Seliger, R. S. Jullens, E. H. Stevens, and I. Lagnado, *J. Vac. Sci. Technol.* **19,** 916 (1981).
7. A. Wagner, *Appl. Phys. Lett.* **40,** 440 (1982).
8. Y. Ochiai, K. Gamo, and S. Namba, *J. Vac. Sci. Technol., B* **1,** 1047 (1983).
9a. M. Kumoro, N. Atoda, and H. Kawakatsu, *J. Electrochem. Soc.* **126,** 483 (1979).
9b. L. Karapiperis and C. A. Lee, *Appl. Phys. Lett.* **35,** 395 (1979).
10. L. W. Swanson, G. A. Schwind, H. E. Bell, and J. E. Brady, *J. Vac. Sci. Technol.* **16,** 1864 (1979).
11. A. Wagner and T. M. Hall, *J. Vac. Sci. Technol.* **16,** 1871 (1979).
12. K. Gamo, T. Ukegawa, Y. Inomoto, Y. Ochiai, and S. Namba, *J. Vac. Sci. Technol.* **19,** 1182 (1981).
13. G. Hanson and B. M. Siegel, *J. Vac. Sci. Technol.* **16,** 1875 (1979).
14. M. P. Lepselter and W. T. Lynch, *in* "VLSI Electronics: Microstructure Science" (N. G. Einspruch, ed.), Vol. 1, p. 83. Academic Press, New York, 1981.
15. R. K. Watts and J. R. Maldonado, *in* "VLSI Electronics: Microstructure Science" (N. G. Einspruch, ed.), Vol. 4, p. 55. Academic Press, New York, 1982.
16. R. Newman, ed., "Fine Line Lithography." North-Holland Publ., Amsterdam, 1980.

17. G. R. Brewer, ed., "Electron-Beam Technology in Microelectronic Fabrication." Academic Press, New York, 1980.
18. I. Brodie and J. J. Muray, "Physics of Microfabrication." Plenum, New York, 1982.
19a. R. L. Seliger and J. W. Ward, *J. Vac. Sci. Technol.* **12,** 1378 (1975).
19b. R. L. Seliger, R. L. Kubena, R. D. Onley, J. W. Ward, and V. Wang, *J. Vac. Sci. Technol.* **16,** 1610 (1979).
19c. J. L. Bartelt, C. W. Slaymon, J. E. Wood, J. V. Chen, C. M. McKenna, C. P. Minning, J. F. Coakley, R. E. Holman, and C. M. Perrygo, *J. Vac. Sci. Technol.* **19,** 1166 (1981).
20. K. Gamo, Y. Inomoto, Y. Ochiai, and S. Namba, *Proc. Int. Conf. Electron, Ion Beam Sci. Technol., 10th* (1982) p. 422.
21. I. Adesida, J. D. Chin, L. Rathburn, and E. D. Wolf, *J. Vac. Sci. Technol.* **21,** 666 (1982).
22a. K. Gamo, Y. Ochiai, and S. Namba, *Jpn. J. Appl. Phys.* **21,** 1 (1982).
22b. Y. Ochiai, K. Gamo, and S. Namba, *J. Vac. Sci. Technol, B* **1,** 1047 (1983).
23a. J. F. Ziegler, *in* "Ion Implantation Science and Technology" (J. F. Ziegler, ed.), p. 51. Academic Press, New York, 1984.
24. J. Lindhard, M. Scharff, and H. E. Schiott, *Mat.-Fys. Medd.—K. Dan. Vidensk. Selsk.* **33,** No. 14 (1963).
25a. J. Lindhard and A. Winther, *Mat.-Fys. Medd.—K. Dan. Vidensk. Selsk.* **34,** No. 4 (1964).
25b. J. Lindhard, V. Nielsen, and M. Scharff, *Mat.-Fys. Medd.—K. Dan. Vidensk. Selsk.* **36,** No. 10 (1968).
25c. S. Lindhard, V. Nielsen, M. Scharff, and P. V. Thomsen, *Mat.-Fys. Medd.—K. Dan. Vidensk. Selsk.* **33,** No. 10 (1968).
26. H. H. Andersen and J. F. Ziegler, "Stopping and Ranges of Ions in Matter. Vol. 3: Hydrogen Stopping Powers and Ranges in All Elements." Pergamon, New York, 1977.
27. J. F. Ziegler, "Helium Stopping Powers and Ranges in All Elements." Pergamon, New York, 1978.
28. J. Lindhard and M. Scharff, *Phys. Rev.* **124,** 128 (1961).
29. W. Brandt and M. Kitagawa, *Phys. Rev. B* **25,** 5631 (1982).
30. J. P. Biersack and L. G. Haggmark, *Nucl. Instrum. Methods* **174,** 257 (1980).
30a. J. F. Ziegler, J. P. Biersack, and U. Littmark, "The Stopping and Range of Ions in Solids," Vol. 1. Pergamon, New York, 1984.
30b. J. P. Biersack and L. G. Haggmark, *Nucl. Instrum. Methods* **174,** 257 (1980).
31. I. Adesida and L. Karapiperis, *Radiat. Eff.* **61,** 223 (1982).
32a. L. Karapiperis, Ph.D. Thesis, Cornell Univ. Ithaca, New York, 1982.
32b. L. Karapiperis, I. Adesida, C. A. Lee, and E. D. Wolf, *J. Vac. Sci. Technol.* **19,** 259 (1981).
33. I. Adesida, C. Anderson, and E. D. Wolf, *J. Vac. Sci. Technol., B* **1,** 1182 (1983).
34. I. Adesida and L. Karapiperis, *J. Appl. Phys.* **56,** 1801 (1984).
35. H. Ryssel, K. Habergers, and H. Kranz, *J. Vac. Sci. Technol.* **19,** 1358 (1981).
36. K. Moriwaki, Ph.D. Thesis, Osaka Univ., Osaka, Japan, 1982.
37a. M. Rosenfield, M.S. Thesis, University of California, Berkeley, 1981.
37b. A. R. Neureuther, D. F. Kyser, and Chiu H. Jing, *IEEE Trans. Electron Devices* **ED-26,** 686 (1979).
38. L. Karapiperis and C. A. Lee, *Appl. Phys. Lett.* **35,** 395 (1979).
39. M. Zhang, I. Adesida, R. Tiberio, and E. D. Wolf, *in* "Microcircuit Engineering '83" (H. Ahmed, J. R. A. Cleaver, and G. A. C. Jones, eds.), p. 157. Academic Press, London, 1983.
40. I. Adesida, E. Kratschmer, E. D. Wolf, A. Muray, and M. Isaacson, *J. Vac. Sci. Technol., B* **3,** 45 (1985).

41a. J. E. Jensen, *Solid State Technol.* **27,** 145 (1984).

41b. R. G. Brault, R. L. Kubena, and J. E. Jensen, *Polym. Eng. Sci.* **23,** 941 (1983).

41c. R. I. Seliger and A. Sullivan, *Electronics* Mar., p. 142 (1980).

42a. T. M. Hall, A. Wagner, and L. F. Thompson, *J. Appl. Phys.* **53,** 3997 (1982).

42b. R. G. Brault and L. J. Miller, *Polym. Eng. Sci.* **20,** 1064 (1980).

42c. M. Mladenov and B. Emooth, *Appl. Phys. Lett.* **38,** 1000 (1981).

43a. Y. Wada, M. Mititaka, K. Mochiji, and H. Obayashi, *J. Electrochem. Soc.* **130,** 187 (1983).

43b. Y. Wada *et al., J. Electrochem. Soc.* **130,** 1127 (1983).

43c. K. Moriwaki, H. Aritome, S. Namba, and L. Karapiperis, *Jpn. J. Appl. Phys.* **20,** 881 (1981).

44a. I. Adesida, J. D. Chin, L. Rathburn, and E. D. Wolf, *J. Vac. Sci. Technol.* **21,** 666 (1982).

44b. A. Wagner, D. Barr, T. Venkatesan, W. S. Crane, V. E. Lambrecti, K. L. Tai, and R. G. Vadinsky, *J. Vac. Sci. Technol.* **19,** 1363 (1981).

44c. A. Muray, M. Scheinfein, M. Isaacson, and I. Adesida, *J. Vac. Sci. Technol., B* **3,** 367 (1985).

44d. W. M. Geis, J. N. Randall, T. F. Deutsch, N. N. Efremow, J. Donnelly, and J. P. Woodhouse, *J. Vac. Sci. Technol,* B **1,** 1178 (1983).

45. D. B. Rensch, R. L. Seliger, G. Csanky, R. D. Olney, and H. L. Stover, *J. Vac. Sci. Technol.* **16,** 1897 (1979).

46. H. Paik, G. N. Lewis, E. J. Kirkland, and B. M. Siegel, *J. Vac. Sci. Technol., B* **3,** 75 (1985).

47a. E. W. Mueller and T. T. Tsong, "Field Ion Microscopy, Principles and Applications." Elsevier, New York, 1969.

47b. H. D. Beckey, "Field Ionization Mass Spectroscopy," Pergamon, New York, 1971.

47c. R. Gomer, "Field Emission and Field Ionization." Harvard Univ. Press, Cambridge, Massachusetts, 1961.

47d. J. J. Hern and S. Ranganathan, eds., "Field Ion Microscopy." Plenum, New York, 1968.

48. S. J. Jason, B. Halpern, M. G. Inghram, and R. Gomer, *J. Chem. Phys.* **52,** 2227 (1970).

49a. R. Levi-Setti, Proc. *Annu. SEM Symp. 7th,* (O. Johari, ed.) p. 125. ITT, Chicago, Illinois, 1974.

49b. W. H. Escovitz, T. R. Fox, and R. Levi-Setti, *IEEE Trans. Nucl. Stud.* **NS-26,** 1147 (1979).

49c. J. Orloff and L. W. Swanson, *J. Vac. Sci. Technol.* **12,** 1209 (1975).

49d. J. Orloff and L. W. Swanson, *J. Appl. Phys.* **50,** 6026 (1979).

50. G. R. Hanson and B. M. Siegel, *J. Vac. Sci. Technol.* **19,** 1176 (1981).

51. J. R. Brady, M.Sc. Thesis, Cornell Univ., Ithaca, New York, 1981.

52a. L. Veneklasen and B. M. Siegel, *J. Appl. Phys.* **43,** 1600 (1972).

52b. H. P. Kuo and B. M. Siegel, *Proc. Int. Conf. Electron Ion Beam Sci. Technol.* (R. Bakish, ed.), Vol. 78-5, p. 3. Electrochem. Soc., Princeton, New Jersey, 1978.

53. L. W. Swanson and L. C. Crouser, *J. Appl. Phys.* **40,** 4741 (1969).

54. R. J. Blackwell, J. A. Kubby, G. N. Lewis, and B. M. Siegel, *J. Vac. Sci. Technol., B* **3,** 82 (1985).

55. P. R. Schwoebel and G. R. Hanson, *J. Appl. Phys.* **56,** 210 (1984).

56a. G. R. Hanson, Report on Gaseous Field Ionization Sources to the National Research and Resource Facility for Submicron Structures (NRRFSS). Cornell Univ., Ithaca, N.Y., 1985.

56b. P. R. Schwoebel and G. R. Hanson, *J. Vac. Sci. Technol., B* **3**, 214 (1985).
56c. P. R. Schwoebel, Ph. D. Thesis, Cornell Univ., Ithaca, N. Y., 1987.
57a. J. A. Kubby and B. M. Siegel, *J. Vac. Sci. Technol. B* **4**, 120 (1986).
57b. J. A. Kubby and B. M. Siegel, *Nucl. Instrum. Methods Phys. Res. Sect. B* **13**, 319 (1986).
58. G. A. M. van Eekelen, *Surf. Sci.* **21**, 21 (1970).
59a. M. Kumoro, H. Hiroshima, H. Tanoue, and T. Kanayama, *J. Vac. Sci. Technol., B* **1**, 985 (1983).
59b. L. R. Harriott, R. E. Scotti, K. D. Cummings, and A. F. Ambrose, *Appl. Phys. Lett.* **48**, 1704 (1986).
60. K. Horiuchi, T. Itakura, and S. Yamamoto, *in* "Microcircuit Engineering '84" (H. Beneking and A. Heuberger, eds.), p. 132.1. Academic Press, London, 1985.
61a. V. E. Krohn, *Prog. Astronaut. Rocketry* **5**, 73 (1961).
61b. V. E. Krohn and G. R. Ringo, *Appl. Phys. Lett.* **27**, 479 (1975).
62. Proceedings of the 1984 International Symposium on Electron, Ion and Photon Beams, *J. Vac. Sci. Technol., B* **3**, 40 (1985); **4**, 60 (1986); **5**, 1 (1987).
63. Proceedings of the International Field Emission Symposium, *J. Phys., Colloq. (Orsay, Fr.)* **C9**, Suppl. to **45**, No. 12 (1984).
64a. MIG300 Metal Ion Gun, VG Scientific, Ltd., East Grinstead, Sussex, England, 1985.
64b. FEI Co., McMinnville, Oregon.
65. G. I. Taylor, *Proc. R. Soc. London, Ser. A* **280**, 383 (1964).
66a. D. R. Kingham, *Appl. Phys. A* **31**, 161 (1983).
66b. D. R. Kingham and L. W. Swanson, *Appl. Phys. A* **34**, 123 (1984).
66c. D. R. Kingham and L. W. Swanson, *Appl. Phys. A* **41**, 157 (1986).
67. L. W. Swanson and D. R. Kingham, *Appl. Phys. A* **41**, 223 (1986).
68. N. Sujatha, P. H. Cutler, E. Kages, J. P. Rogers, and N. M. Miskovsky, *J. Appl. Phys.* **A32**, 55 (1983).
69. R. Gomer, *J. Appl. Phys.* **19**, 365 (1979).
70. P. D. Prewett, G. L. R. Mair, and S. P. Thompson, *J. Phys. D* **15**, 1339 (1982).
71. J. W. Ward and R. L. Seliger, *J. Vac. Sci. Technol.* **19**, 1186 (1981).
72. R. C. Forbes and G. L. R. Mair, *J. Phys. D* **15**, L153 (1982).
73a. Tiven Katesan, A. Wagner, and D. Barr, *Appl. Phys. Lett.* **38**, 943 (1981).
73b. G. L. R. Mair and A. von Engel, *J. Phys. D* **14**, 1721 (1981).
73c. G. L. R. Mair, *J. Phys. D* **15**, 2523 (1982).
74. H. Gaubi, P. Sudraud, M. Tence, and J. Van de Walle, *Proc. Int. Field Emiss. Symp. 29th, Goeteborg, Swed.* (1982).
75. P. Givet, "Electron Optics" (P. W. Hawkes, transl.), 2nd Engl. Ed. Pergamon, New York, 1972.
76. A. B. El-Kareh and J. C. El-Kareh, "Electron Beams, Lenses and Optics," Vols. 1 and 2, Academic Press, New York, 1970.
77a. A. Septier, ed., "Focusing of Charged Particles," Vols. I and II. Academic Press, New York, 1967.
77b. A. Septier, ed., "Applied Charged Particle Optics," Advances in Electronics and Electron Physics, Suppls. 13A and B. Academic Press, New York, 1980.
78. E. Harting and F. H. Read, "Electrostatic Lenses." Elsevier, New York, 1976.
79a. E. Munro, *in* "Image Processing and Computer Aided Design in Electron Optics" (P. W. Hawkes, ed.), p. 284. Academic Press, New York, 1973.
79b. H. C. Chu and E. Munro, *Optik (Stuttgart)* **61**, 121 (1981).
80. P. W. Hawkes, ed., "Image Processing and Computer-Aided Design in Electron Optics." Academic Press, New York, 1973.

81a. J. C. Eidson and R. K. Scudder, *J. Vac. Sci. Technol.* **19**, 932 (1981).
81b. J. Kelly, T. Groves, and H. P. Kuo, *J. Vac. Sci. Technol.* **19**, 936 (1981).
82.   L. H. Veneklasen, *J. Vac. Sci. Technol., B* **3**, 185 (1985).
83a. R. D. Moore, G. A. Caccoma, H. C. Pfeiffer, E. V. Weber, and O. C. Woodward, *J. Vac. Sci. Technol.* **19**, 950 (1981).
83b. P. J. Coane, D. P. Kern, A. J. Speth, and T. H. P. Chang, *Proc. Conf. Electron Ion Beam Sci. Technol., 10th* (R. Bakish, ed.), Vol. 83-2, p. 2. Electrochem. Soc., Princeton, New Jersey, 1983.
84.   G. H. N. Riddle, *J. Vac. Sci. Technol.* **15**, 857 (1978).
85.   J. Orloff and L. W. Swanson, *in* "Scanning Electron Microscopy 1979" (O. Johari, ed.), p. 39. AMF, O'Hare, Illinois, 1979.
86.   H. Paik, G. N. Lewis, E. J. Kirkland, and B. M. Siegel, *J. Vac. Sci. Technol., B* **3**, 75 (1985).
87a. M. Szilagyi, H. Paik, and B. M. Siegel, *Proc. Int. Conf. Electron Ion Beams Sci. Technol., 10th* (R. Bakish, ed.), Vol. 83-2, p. 409. Electrochem. Soc., Princeton, New Jersey, 1983.
87b. M. Szilagyi, *J. Vac. Sci. Technol., B* **1**, 1137 (1983).
88.   K. Kurihara, *J. Vac. Sci. Technol., B* **3**, 41 (1985).
89.   E. Munro, *J. Vac. Sci. Technol.* **12**, 1146 (1975).
90a. H. Ohiwa, *J. Phys. D* **10**, 1437 (1977).
90b. H. Ohiwa, *Ultramicroscopy* **15**, 221 (1984).
91.   H. Paik and B. M. Siegel, *Optik (Stuttgart)* **70**, 152 (1985).
92.   J. Kelly, *Adv. Electron. Electron Phys.* **43**, 43 (1977).
93a. G. N. Lewis, H. Paik, J. Mioduszewski, and B. M. Siegel, *J. Vac. Sci. Technol. B* **4**, 116 (1986).
93b. G. N. Lewis, E. J. Kirkland, J. Mioduszewski, D. Weiner, and B. M. Siegel, submitted *J. Vac. Sci. Technol.* 1987.
94a. R. Levi-Setti, X. L. Wong, and G. Crow, *J. Phys. (Orsay, Fr.)* **45**, C9-197 (1984).
94b. T. Shiokawa, P. H. Kim, K. Toyoda, S. Namba, T. Matsui, and K. Gamo, *J. Vac. Sci. Technol., B* **1**, 1117 (1983).
94c. M. Kumoro, *Thin Film Solids* **92**, 155 (1982).
94d. M. Kumoro and H. Kawakatsu, *J. Appl. Phys.* **52**, 2642 (1981).
94e. T. Tsumagari, H. Ohiwa, T. Okutani, and T. Noda, *J. Vac. Sci. Technol., B* **1**, 1121 (1983).
94f. E. Miyauchi, H. Arimoto, H. Hashimoto, and T. Utsumi, *J. Vac. Sci. Technol., B* **1**, 1113 (1983).
94g. T. Kato, H. Morimoto, K. Saitoh, and H. Nakata, *J. Vac. Sci. Technol., B* **3**, 50 (1985).
94h. J. R. A. Cleaver, P. J. Heard, and H. Ahmed, *in* "Microcircuit Engineering '83" (H. Ahmed, J. R. A. Cleaver, and G. A. C. Jones, eds.), p. 135. Academic Press, New York, 1983.
95.   J. L. Bartelt, C. W. Slayman, J. E. Wood, J. Y. Chen, C. M. McKennon, C. P. Minning, J. F. Coakley, R. F. Holman, and C. M. Perrygo, *J. Vac. Sci. Technol.* **19**, 1166 (1981).
96a. N. P. Economou, D. C. Flanders, and J. P. Donnelly, *J. Vac. Sci. Technol.* **19**, 1172 (1981).
96b. J. N. Randall, D. C. Flanders, N. P. Economou, J. P. Donnelly, and E. L. Bromley, *J. Vac. Sci. Technol., B* **1**, 1152 (1983).
96c. J. N. Randall, D. C. Flanders, N. P. Economou, J. P. Donnelly, and E. L. Bromley, *J. Vac. Sci. Technol., B* **3**, 58 (1985).

10

97a. G. Stengl, R. Kaitna, H. Loschner, P. Wolf, and R. Sacher, *J. Vac. Sci. Technol.* **16,** 1883 (1979).

97b. G. Stengl, R. Kaitna, H. Loschner, R. Reider, P. Wolf, and R. Sacher, *J. Vac. Sci. Technol.* **19,** 1164 (1981).

98a. G. Stengl, H. Loschner, W. Maurer, and P. Wolf, *Proc. Symp. Submicron Microlithogr.* SPIE, Bellingham, Washington, 1985.

98b. G. Stengl, H. Loschner, W. Maurer, and P. Wolf, *J. Vac. Sci. Technol. B* **4,** 194 (1986).

99. W. L. Brown, T. Venkatesan, and A. Wagner, *Solid State Technol.* **24,** 60 (1981).

100. H. P. Kuo, J. Foster, W. Haase, J. Kelly, and B. M. Oliver, *Proc. Int. Conf. Electron Ion Beam Sci. Technol., 10th* (R. Bakish, ed.), Vol. 83-2, p. 78. Electrochem. Soc., Princeton, New Jersey, 1983.

101. H. Paik, E. J. Kirkland, and B. M. Siegel, *J. Phys. E* **20,** 61 (1987).

102. H. Paik, Ph.D. Thesis, Chaps. VIII and IX. Cornell Univ., Ithaca, New York, 1985.

103. H. Boersch, *Z. Phys.* **139,** 115 (1954).

104. F. Lenz, *Proc. Int. Conf. Electron Microsc., 4th* p. 39 (1958).

105. H. Rose and R. Spehr, *Adv. Electron. Electron Phys., Suppl.* **13C,** 475 (1983).

106. H. Rose and R. Spehr, *Optik (Stuttgart)* **57,** 339 (1980).

107. K. H. Loeffler, *Z. Angew. Phys.* **27,** 145 (1969).

108a. W. Knauer, *Optik (Stuttgart)* **54,** 211 (1979).

108b. W. Knauer, *Optik (Stuttgart)* **59,** 335 (1981).

109. A. V. Crewe, *Optik (Stuttgart)* **50,** 205 (1978).

110. A. B. El-Kareh and M. A. Smithers, *J. Appl. Phys.* **50,** 5596 (1979).

111a. T. Sasaki, *Proc. VLSI Conf.* p. 125 (1979).

111b. T. Sasaki, *Proc. Conf. Electron Ion Laser Beam Technol., 9th* p. 73 (1980).

112. T. Groves, D. L. Hammond, and H. P. Kuo, *J. Vac. Sci. Technol.* **16,** 1680 (1979).

113. Y. W. Yau, T. R. Groves, and R. F. W. Pease, *J. Vac. Sci. Technol., B* **1,** (1983).

114. R. E. Howard, P. F. Liao, W. J. Skocpol, L. D. Jackel, and H. G. Craighead, *Science* **221,** 117 (1983).

115. S. Mackie and S. Beaumont, *Solid State Technol.* **28,** 117 (1985).

116. P. M. Mankiewich, R. E. Howard, L. D. Jackel, W. J. Skocpol, and D. M. Tennant, *J. Vac. Sci. Technol. B* **4,** 380 (1986).

117. P. Santhanam, S. Wind, and D. E. Prober, *Phys. Rev. Lett.* **53,** 1179 (1984).

118. A. C. Warren, D. A. Antoniadis, H. I. Smith, and J. Melngailis, *J. Vac. Sci. Technol. B* **4,** 365 (1986).

119. H. Smith, *J. Vac. Sci. Technol. B* **4,** 148 (1986).

Chapter **6**

Alignment Techniques in Optical and X-Ray Lithography

M. FELDMAN

AT&T Bell Laboratories
Murray Hill, New Jersey 07974

I. INTRODUCTION

There is no best method of optical alignment.

Whatever advantages one method of alignment may have over another are not as important as the differences in engineering used to construct a

229

practical system: A well-engineered version of a relatively poor method is far better than the reverse. A consequence of this is that different alignment systems abound, and there is little correlation between how successful they are and how cleverly the alignment is done.

Rather than categorizing the alignment systems in use today, this chapter summarizes the basic principles important for alignment and states their advantages and disadvantages. Many systems are based on several of these principles. The hope is that an understanding of the underlying principles will enable the reader to appreciate current systems and to follow developments in this evolving field.

The overwhelming majority of alignment techniques are optical, and we restrict the discussion of alignment to optical methods. However, for reference, a brief discussion of electrical methods used to characterize previous alignments is also given.

II. REVIEW OF OPTICAL PRINCIPLES

A. Lenses and Zone Plates

A *lens* can be defined as an optical element that forms a pointlike image of a distant point [1], that is, all of the rays that leave the object point in Fig. 1a and enter the lens end up at the image point. (For our purposes the presence of reflecting elements in the lens is irrelevant.) A necessary and sufficient condition for good *diffraction-limited* imaging is that all of the rays have the same total optical path length, so that the number of wavelengths separating the image and the source is independent of the ray path, to within an accuracy of $\lambda/4$ (λ is the wavelength; it is reduced by a factor of n, the index of refraction, within a medium). Practical lenses are required to remain diffraction limited over a *field* of object points and an often quite narrow range of wavelengths.

Discontinuities in the lens may introduce changes in the optical path length between adjacent rays. If these changes are a multiple of λ, the imaging remains diffraction limited. Such an arrangement is a *Fresnel lens* (Fig. 1b); usually, however, Fresnel lenses are used in applications that are far from diffraction limited.

In a *phase-type Fresnel zone plate,* a discontinuity occurs at every $\lambda/2$ in optical path length (Fig. 1c). This divides the area of the zone plate into symmetrically located zones, with an optical thickness difference of $\lambda/2$ between adjacent zones. No variation in optical thickness occurs within a zone. Nevertheless, since the maximum departure from equal (within a

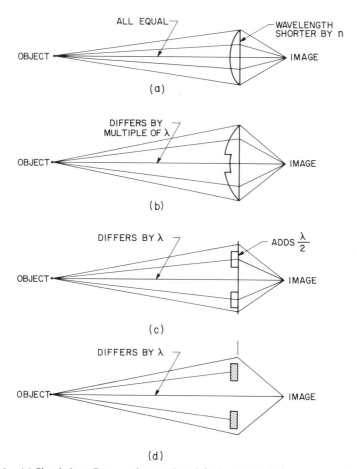

Fig. 1. (a) Simple lens. Because the wavelength is shortened within the lens material, the total number of wavelengths in all the optical paths are equal. (b) Fresnel lens. The total number of wavelengths in rays through one section of the lens differs by a multiple of the wavelength from rays through other sections. (c) Phase-type Fresnel zone plate. Adjacent zones differ by a constant half-wavelength. (d) Amplitude-type Fresnel zone plate. Adjacent zones are opaque.

multiple of λ) total optical path length for all rays is $\lambda/4$, the imaging is diffraction limited.

Unlike a lens, a zone plate does not direct all of the incident light to the image; a virtual image that is approximately equally intense is formed at the negative focal plane (for a distant object) as are numerous symmetrically placed faint "higher-order" images. The maximum efficiency (fraction of light collected in one image) for a practical phase zone plate is 40%.

If the optical path length difference between adjacent zones differs from $\lambda/2$, the efficiency is lower, nominally falling to zero at integral multiples (including 0) of λ.

In an *amplitude zone plate* (Fig. 1d), alternate zones are opaque (or at least gray). At maximum contrast the efficiency is $\sim 10\%$. For a high-contrast amplitude zone plate the phase of adjacent zones is irrelevant; however, practical zone plates may contain a combination of phase and amplitude differences between zones. The ability to pattern zone plates, with their imaging properties, on masks and wafers leads to an important class of alignment techniques.

B. Numerical Aperture

Lenses (and zone plates) are characterized by their focal lengths f and their f/numbers F or numerical apertures (NA) [1]. The f/number and NA of a lens are defined in Fig. 2 for an object in air. They are related by

$$NA \sqrt{4F^2 + 1} = 1 \tag{1}$$

or

$$2 \, NA \times F \approx 1 \qquad \text{for large } F. \tag{2}$$

In a medium of refractive index n, $NA = n \sin \theta$.

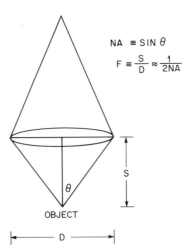

$$NA \equiv SIN \, \theta$$
$$F \equiv \frac{S}{D} \approx \frac{1}{2NA}$$

Fig. 2. Definitions of the numerical aperture (NA) and f/number (F) of a lens. The longer conjugate is often assumed to be infinite.

1. Resolution

As a rule of thumb, the resolution of a diffraction-limited lens is $\sim \lambda F$, or its rough equivalent $\lambda/2$ NA. A more precise measure of the resolution is the *modulation transfer function* (MTF), or response of the lens to a family of gratings whose spatial frequency varies from zero to the maximum that the lens can image. The spatial frequency is typically measured in cycles per millimeter or line pairs per millimeter and corresponds to half the number of "lines" or elements imaged per millimeter.

The MTF of a diffraction-limited lens is shown in Fig. 3 as a function of the spatial frequency of the gratings imaged. Two cases are illustrated:

(1) In incoherent illumination, light from the object is directed to all parts of the lens, completely filling the entrance aperture of the lens.

(2) In coherent illumination, light from the object is directed only at the center of the lens. If not for the role played by diffraction at the object, all but the central portion of the lens would be unilluminated and therefore wasted!

The difference in response due to illumination can be seen from the figure. For coherent illumination, the lens retains full response up to a spatial frequency of NA/λ and then has zero response for higher frequencies. For incoherent illumination, the lens response extends to twice as high a spatial frequency, 2 NA/λ, but at the price of reduced MTF throughout. Microscopes are generally operated with incoherent illumination *(Köhler*

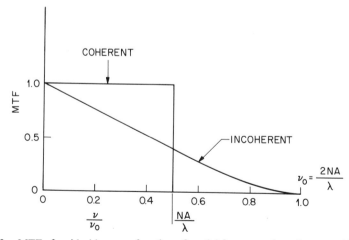

Fig. 3. MTF of an ideal lens as a function of spatial frequency, for coherent and incoherent illumination.

illumination) because the eye does not need much contrast to see, and the higher the resolution in a microscope the better. Lithographic lenses, on the other hand, are usually operated rather closer to coherent than to incoherent illumination. Note that the rule of thumb resolution λF, or $\lambda/2$ NA, corresponds to the spatial frequency NA$/\lambda$, or halfway through the range of spatial frequencies that produce any modulation.

2. Depth of Focus

The size of the image of a point object increases hyperbolically from its minimum, or waist, at focus. Far from focus, the asymptotes of the blur are defined by the f/number of the optical cone. It is customary to define the depth of focus as that position, on either side of focus, at which the asymptotes have reached twice the size of the focused waist. In Fig. 4 we have

$$x = z/F = 2\lambda F, \tag{3}$$

$$z = 2\lambda F^2 \approx \lambda/2 \text{ NA}^2. \tag{4}$$

Note that the *range* of focus is $2z$, or λ/NA^2. The resolution $\lambda F = x/2$ can also be expressed as a function of the depth of focus z:

$$\text{Resolution} = \sqrt{\lambda z/2}. \tag{5}$$

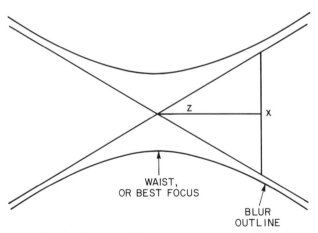

WAIST,
OR BEST FOCUS

BLUR
OUTLINE

Fig. 4. Diameter of the image of a point, near focus.

C. Illumination

1. Bright Field versus Dark Field

Most commonly, objects that are viewed by an optical system are illuminated by a symmetric cone of light whose central, or *chief* ray passes through the center of the entrance aperture of the optics (Fig. 5a). Featureless objects, either in transmission or in reflection, appear uniformly bright, since only absorption contributes to loss of light. The edges of objects may scatter some of the light out of the entrance aperture, causing them to appear relatively dark. This method of illumination is called *bright field,* since small objects, which scatter light, generally appear dark against a bright background.

In contrast, dark-field illumination (Fig. 5b) is directed so that no light reaches the entrance aperture of the optics unless it is scattered by the object. Featureless objects thus appear dark, and small objects and feature edges are bright against a dark background.

Dark field is one of several methods of obtaining high contrast by controlling which light striking the object is used in constructing an image. The high contrast is obtained at the cost of an appreciable loss in intensity. For many objects, especially the edges of thin films such as chrome patterns on masks, the resolvable width of the observed image is limited only by the resolution of the optical system.

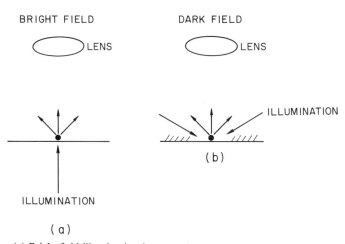

Fig. 5. (a) Bright-field illumination in transmitted light. (b) Dark-field illumination with top lighting.

2. *Image Shifting*

By definition, alignment is concerned with determining the positions of centroids and edges of features on masks and wafers. If both the illumination and the feature profiles are symmetric, the correct centroid position

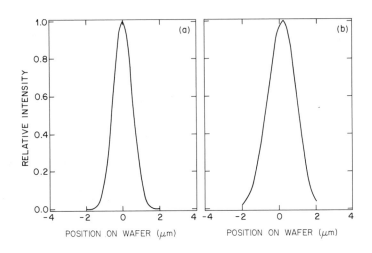

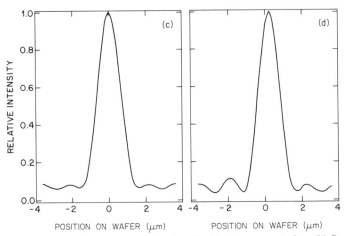

Fig. 6. (a) Video trace of a 1.5-μm line viewed with 1.5-μm resolution. (b) Same as (a) except that the right-hand half of the line is twice as bright as the left-hand half. (c) Video trace from a focused linear zone plate whose minimum dimension is 1.5 μm. (d) Same as (c) except that the rightmost 0.75 μm of each zone is twice as bright as the rest.

will be observed, and edges will be displaced symmetrically (if at all) from their nominal positions. Unfortunately, this is not always the case:

(1) The illumination may be brighter on one side than on the other.

(2) The illumination may cover a different range of angles on one side than on the other.

(3) The feature profile may not be symmetric. In practice this occurs most commonly because of uneven flow of resist over the feature, due to either a striation in the resist or a wake arising from the feature itself.

The shifting of edges, in either bright- or dark-field illumination, depends in detail on the edge and illumination intensity profiles. The effect may well be a substantial fraction of the resist thickness and consequently may be a major contributor to alignment errors.

The shifting of the centroid may be accompanied by a loss of resolution. In Fig. 6a, the response of a 1.5-μm line is shown to an assumed 1.5 μm FWHM (full width at half-maximum) Gaussian resolution. In Fig. 6b, the right-hand 0.75 μm of the line has been arbitrarily increased in brightness by a factor of 2. Not only is the centroid of the response function shifted by the same amount as the center of gravity of the intensity distribution, but the response function is substantially wider. Figure 6c shows the response function of a linear zone plate with minimum linewidth 1.5 μm. When the right-hand 0.75 μm of each zone is increased in intensity by a factor of 2 (Fig. 6d), the centroid shifts, but there is no accompanying loss in resolution. The conclusion is that zone plates are subject to the same kinds of centroid shifts as conventional marks due to uneven brightness, but at least the resolution of the zone plates is less affected.

III. METHODS OF ALIGNMENT

A. Overview of Printers

1. Contact and Proximity Printers

Excellent resolution can be obtained with the simplest of lithographic tools, a contact printer [2]. If an initially flat mask and wafer are brought into intimate contact by evacuation of the space between them and illuminated normally with parallel light, the only source of image degradation is diffraction spreading through the thickness of the resist. In practice gaps of a few micrometers remain in some areas between the mask and the wafer. Index-matching liquid gates have been used, at least in the laboratory, to reduce diffraction spreading across the gaps by reducing λ by a factor of n, the index of refraction of the liquid. The edge blurring (and roughly the minimum resolvable feature size) that can be obtained is given by

$$\delta \approx \sqrt{2\lambda d}, \tag{6}$$

where the variables are defined in Fig. 7 and λ is the wavelength in the medium through which the radiation passes.

Equation (6) also applies to portable conformal masks (PCMs), which are the upper patterned levels of some multilevel resist systems. The PCM is used to define underlying levels with a flood exposure of parallel light. We will not discuss PCMs further, since alignment is not possible for the flood exposure (except for slight displacements by changing the angle of illumination).

Contact printers are hard on masks. Entrapped dirt particles, bits of adhering resist, and silicon particles accumulate, necessitating periodic cleaning or replacement of the masks. Consequently, contact printing is more useful on small chips, where good yields can be obtained in spite of relatively high defect densities.

Alignment is physically impossible during *hard* contact. Therefore, the mask and wafer are aligned while slightly separated, and then clamped together. The separation during alignment is a trade-off between a gap large enough to minimize damage from entrapped particles and small enough for simultaneous focus on both mask and wafer [see Eq. (5)]. In addition, there is some sideways motion during the clamping process, which generally gets worse as the alignment gap is increased.

In proximity printing the final clamping motion is omitted (or decreased), reducing mask damage at the expense of resolution [Eq. (6)]. Many schemes exist for optimizing this trade-off by measuring the gap or the force between the mask and the wafer.

In a particular case of proximity printing the imaging is done with soft

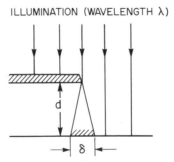

ILLUMINATION (WAVELENGTH λ)

Fig. 7. Diffraction spreading at the edge of a shadow. In a medium of index of refraction n, the wavelength is a factor of n smaller than in vacuum.

x rays, either in a nearly parallel beam from a synchrotron or a storage ring or in a diverging beam from a "point" electron-bombarded or plasma source [3]. For wavelengths of a few angstroms, the diffraction spreading may be negligible compared with the combined effects of penumbral blurring and finite range of the Compton electrons with which the x rays interact. For the purposes of optical alignment x-ray lithography differs from optical lithography in two ways:

(1)   The x-ray mask is generally less transparent and sometimes less flat than the optical mask.

(2)   If a point of x rays is used, the diverging x rays cast a magnified shadow of the mask on the wafer. The increase in magnification, or "runout," is

$$D - d = g(D/L), \tag{7}$$

where the variables are defined in Fig. 8.

2. Projection Printers

Available projection printers have usually either $1:1$ magnification or $5:1$ or $10:1$ reduction. The higher reduction printers *step and repeat* a relatively small imaged field across the wafer. Lenses are not available to image entire 100- or 125-mm wafers with adequate resolution; $1:1$ projection printers either image a small field similar to the reduction printers or image a narrow, curved field that is scanned across the wafer [2].

Projection printers provide long mask life, especially if the mask or

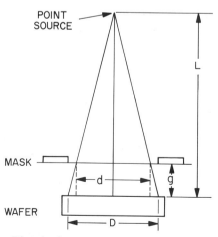

Fig. 8.   Diverging beam geometry used in some x-ray tools.

reticle (single field mask for a step-and-repeat camera) is not frequently changed. High mask quality is particularly important for step-and-repeat cameras, which image only one or a few chips at a time. A fatal defect here may significantly affect the total yield. It is not necessary for 1 : 1 scanning cameras to have perfect masks in order to give statistically good yields, but because their images are not reduced, the linewidth requirements on their masks are correspondingly tighter.

3. Alignment

Alignment is the process of positioning an alignment mark on one circuit level with a corresponding mark on another level. Perfect alignment does not guarantee perfect overlay accuracy, or registration, which are terms that apply to relative position throughout the field. Pattern placement errors including magnification errors and mask distortion affect overlay accuracy, even in the presence of perfect alignment [4].

Often, magnification differences between mask and wafer dominate the distortion. These may arise from differences in temperature between when the mask or wafer was patterned and when it is used or from wafer stresses arising from processing steps. Nonlinear inplane distortions can be minimized by careful wafer processing and by ensuring that both mask and wafer are held flat in their chucks.

In general contact printers do not correct for inplane distortions, severely limiting their applicability for fine-line applications. However, careful control of the mask and wafer temperature can minimize magnification errors, and, in principle, controlled out-of-plane bending of the mask and wafer can provide different magnifications in the X and Y directions.

With the exception of x-ray tools, proximity printers have the same limited control over distortions as contact printers. For x-ray proximity printers, the divergence of the x rays may be exploited to correct the magnification by adjusting the gap between the mask and the wafer. In some methods of zone plate alignment the correct gap may be found as part of the normal alignment procedure; however, it is also feasible to determine the correct gap for a particular mask and wafer lot by careful mechanical control or measurement of the gap or simply by trial and error.

One-to-one projection printers that scan a narrow curved field across the wafer can have programmed magnification corrections. Along the direction of the field small optical changes can be used to change the magnification slightly from unity. Perpendicular to the field, along the direction of scan, mechanical arrangements can be used to scan image and object at slightly different speeds, again resulting in magnification slightly different from unity. Since these adjustments are independent, magnification can simultaneously be optimized in two orthogonal directions.

Step-and-repeat cameras in principle can locate the center of each field at an optimal position on the wafer. This not only minimizes overlay errors due to wafer magnification, but also corrects for other inplane wafer distortions such as shear. However, magnification errors *within* a field are not corrected, since it is not practical to change the magnification of present optical systems at each exposed field. Nevertheless, alignment at each field of a step-and-repeat camera provides the best overlay accuracy obtainable in optical lithography, and alignment at each field of a step-and-repeat x-ray proximity printer provides the best overlay accuracy obtainable in x-ray lithography.

In less critical applications, "global" alignment, or mapping, can be used in step-and-repeat cameras, in which alignment is performed at only a few sites on the wafer. A computer analysis estimates the correct wafer positions for each field, and a highly accurate table (usually interferometer controlled) positions the wafer. Depending on how many sites are measured, magnification and various inplane distortions are corrected with minimal time spent in alignment.

4. Error Budgets

The integrated circuit designer wishes to have one level aligned with a previous one (not necessarily the immediate predecessor) within an alignment tolerance σ. There is no industry rule for specifying alignment accuracy; we shall use 2σ, where σ is the standard deviation of the alignment error; $\sim 95\%$ of all alignments are within $\pm 2\sigma$.

Sometimes several levels are aligned to the same previous level; this may even be a 0-order level containing only alignment marks. If σ_1 and σ_2 are the alignment errors of levels 1 and 2, respectively, to a common mark, the relative alignment between levels 1 and 2 is given by

$$\sigma_{12}^2 = \sigma_1^2 + \sigma_2^2. \tag{8}$$

In some printers, particulary step-and-repeat tools, it is convenient to align both reticle and wafer to the tool, rather than to each other. In this case the relative alignment between the image and the wafer is given by

$$\sigma^2 = (\sigma_r/M)^2 + \sigma_w^2, \tag{9}$$

where σ_r and σ_w are the reticle and wafer errors and M the magnification. If M is a large number, then $\sigma \approx \sigma_w$ even if σ_r is larger than σ_w. In other words, it is sufficient to align the wafer to the reduction camera, as long as the reduction ratio is high and the reticle alignment is not too bad.

Similarly, the errors in overlay accuracy, or registration, which include distortion effects, can be combined between different levels.

B. Capture Range

The capture range, which is the range over which the alignment system will work, may be a critical factor in practical exposure tools. There are two cases:

(1) At alignment the mask and wafer are in the center of the capture range. An example would be a step-and-repeat camera where both reticle and wafer are aligned to the camera. Since optimal performance of the alignment system is required only near the center of the capture range, the requirements on the detection system are relaxed.

(2) Alignment of mask and wafer can be performed anywhere within the capture range. The detection system must have equivalent sensitivity over the entire range.

Since one mask typically exposes many wafers, the starting point for alignment is often a reference to the wafer edge or center. Mechanical tolerances are typically of the order of 10 or 20 μm. In addition, different exposure tools can use different reference surfaces, compounding the problem if previous levels were printed with another type of printer. Often an initial coarse alignment is performed, either manually or automatically. The capture range for the following fine-alignment step is then the alignment accuracy achieved by the coarse alignment.

C. Bifocal Lenses

In both contact and proximity printers the mask and wafer are separated by a finite gap during the alignment process. This gap may be as large as several tens of micrometers for x-ray proximity printers. From Eq. (6) we see that the resolution obtained during alignment is limited by the depth of focus required to view both mask and wafer.

Mechanical means can be used to shift the microscope focus between mask and wafer. For example, one can vibrate the microscope objective vertically or periodically interpose an optical flat, which has the effect of shifting the focal distance. This approach is somewhat awkward; in addition one must be careful to avoid any sideways shift of the image along with the vertical shift in focus.

Alternatively, the microscope objective can be designed so that the mask is in focus in one kind of light at the same time that the wafer is in focus in another. Objectives have been designed with high chromatic abberation [5], so that shifts of up to 100 μm can be obtained over the visible spectrum; in the right wavelengths sharp focus is obtained simultaneously

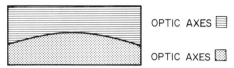

Fig. 9. Bifocal lens element.

at mask and wafer. Unfortunately, out-of-focus images also occur due to the "wrong" wavelengths at mask and wafer. It is important to use a high enough numerical aperture that these images are far out of focus, only degrading the contrast but not confusing the correct images.

Polarization effects are also used to obtain separate sharp images of the mask and wafer [6]. If the plate shown in Fig. 9, made from birefringent material with crossed optic axes, is inserted in the microscope objective, it will act as a weak positive lens for light having one direction of polarization and as a weak negative lens for the other. For crystalline quartz the parameters needed to obtain focal separations of a few tens of micrometers are readily obtainable. It is important that such an element be carefully centered in the microscope objective to prevent sideways shifts of the images. In addition, if the element is located at the back focal plane of the microscope objective, the magnification of the two images will be the same.

D. Contrast

The general rule is the higher the contrast, the better the alignment will be, for both manual and automatic alignment. Some caveats apply:

(1) The eye is most sensitive to abrupt changes in intensity at feature edges. Hence, for manual alignment some differentiation (or the equivalent) of the image may be better than a simple increase in contrast.

(2) High contrast may imply loss of information. For example, an edge may be located by observing when a threshold from dark to light is crossed, but a computer fit to several points near the threshold will average the noise and provide a more accurate answer.

(3) For an automatic system, knowledge of the threshold is critical. For example, adding a uniform illumination to the entire alignment region will degrade the contrast, but a change in threshold will restore the original signal (apart from possible increases in system noise).

Because of the importance of high contrast in later image processing, a number of techniques have evolved to enhance the contrast of alignment marks:

(1) *Placement of a dark mark on the wafer by anisotropic etching.* For example, KOH is used to etch 70° triangular grooves in 100-plane silicon wafers. When illuminated by nearly vertical light, these grooves appear dark, because after reflections from the two sides the light is at an angle that misses the microscope objective.

(2) *Dark field.* As discussed above, light incident at a steep angle illuminates the edges of marks on the mask and wafer, making these appear bright against a dark background. As noted, the illumination must be uniform and the edge profiles of the marks symmetric to prevent shifting of the images.

(3) *Diffractive coding.* The mark is made periodic, so that it will diffract light in particular directions. The viewing is then restricted to these directions so that virtually all of the light that is detected has indeed been diffracted by the mark. For example, a dotted line can be used to determine a location in a direction perpendicular to its length, with the viewing contrast enhanced by having only light scattered along the direction of the line admitted by the lens system aperture. One simple arrangement is shown in Fig. 10.

(4) *Light concentration.* Circular Fresnel zone plates can be used to focus a laser beam to a diffraction-limited spot. Even with diffraction efficiencies that may be of the order of 5% the decrease in area due to the focusing is so large that the contrast between the intensity of the focused spot and its surrounding background may exceed 100 : 1.

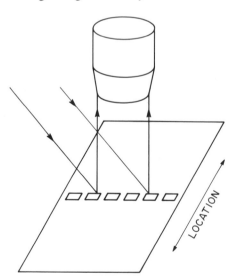

Fig. 10. Use of diffraction to enhance contrast. Only diffracted light enters the microscope objective. The high contrast with which the line is viewed improves the determination of its location.

E. Manual Alignment Methods

The eye is most sensitive to detecting asymmetry in a pattern and detecting rapid changes in brightness at edges. We shall discuss manual alignment techniques that exploit these sensitivities.

1. Symmetric Patterns

Alignment consists of superimposing patterns on the mask and wafer such that they appear to be symmetrically located. The prototype of this kind of alignment is centering one pattern, typically a square, within another larger outline (Fig. 11a). Similarly, a narrow opaque cross may be centered within a wider hollow cross, providing the operator with more areas to check for symmetry (Fig. 11b).

These simple patterns can be expanded to achieve a kind of vernier effect: The squares are surrounded by a field of squares in which there are programmed offsets (Fig. 11c). Thus, when the central square is centered, the squares on the left will be offset by an amount that is equal and opposite to the offset in the squares on the right, etc. This again increases the order of the symmetry that the operator strives to obtain.

A similar effect is achieved by introducing a small slant into feature edges. For example, if the arms of the central cross in Fig. 11b are tapered, the effect of a misalignment is an exaggerated displacement between the intersections of the opposite sides of the arms (Fig. 11d). Unfortunately, these small angles are difficult to achieve in practice, since most pattern generators use a square grid address structure.

The performance of manual alignment using these symmetric alignment patterns is determined largely by the resolution and contrast of the viewed images. Although more complex patterns may appear to offer advantages, they probably do not help unless the image the operator sees is similarly improved.

2. Matching Edges

The eye is capable of accurately aligning two line segments so that they are collinear. It is important that the edges of the lines be sharply defined; the widths are much less critical. A short separation between the ends is also helpful.

Quite separate line segments can be used for alignment in the X and Y directions (Fig. 12a); equivalently, the segments can be combined into a cross or right angle (Fig. 12b). In one arrangement that has been used the tone of one segment is opposite to that of the other; thus, matching the ends takes on some of the elements of a symmetry alignment (Fig. 12c).

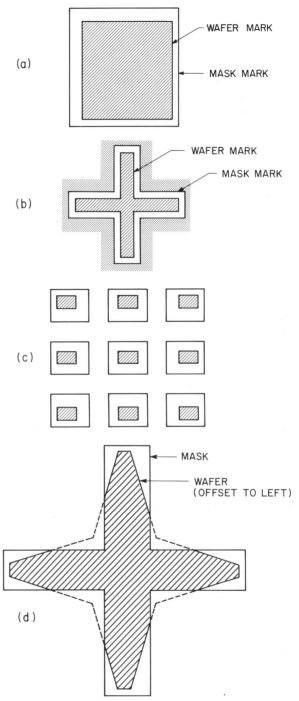

Fig. 11. (a) Square within a square alignment mark. (b) Cross within a cross alignment mark. (c) Array of squares alignment mark. (d) Tapered cross within a cross alignment mark.

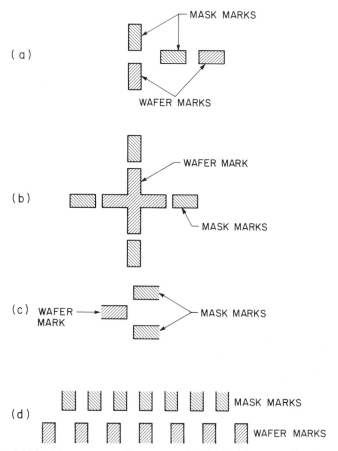

Fig. 12. (a) Matching segments alignment mark. (b) Matching cross alignment mark. (c) Matching segments of opposite tone alignment mark. (d) Vernier alignment mark.

Arrays of segments can also be matched. By using a slightly different spacing between the segments, a vernier is obtained, and alignment consists not only of matching the right segments, but also of observing the symmetric offsets in adjacent line segments (Fig. 12d). A particular advantage of vernier manual alignments is that calibrated offsets are readily obtained; this is difficult with other manual techniques.

Manual alignment is also accomplished by superimposing similar periodic patterns from the mask and wafer, which have slightly different periods. The Moire fringes that are obtained are much larger than the individual lines and are readily viewed. In fact, it is desirable that the viewing optics have limited resolution, so that only the Moire fringes and *not* the individual lines are resolved. As with many systems using periodic structures, there may be an ambiguity: The alignment may be in error by one period (i.e., fringe). However, with at least one Moire pattern, circular lines and spaces, a misalignment of one period is obvious [7].

It should be emphasized that for all of the manual alignment systems discussed above performance is dominated by the contrast and resolution of the images the operator views. The particular methods and patterns used vary widely, but each depends on human judgment to accomplish the alignment, and their ultimate capabilities are rather similar.

F. Automatic Alignment Methods

1. Average Transmission

In the conceptually simplest method of automatic alignment a mark on the wafer is viewed through a window of the same width on the mask (Fig. 13a). The viewing takes place via an imaging system, for example, the primary lens in a projection printer, or simple shadowing, as in a proximity printer. Suppose the mark is darker than the area on the wafer surrounding it. Then the total amount of light reflected from the wafer and passing through the mask will be at a minimum at alignment.

This minimum is searched for in a number of ways:

(1) The wafer can be slowly scanned with respect to the mask and the total light detected.

(2) The image of the window can be moved on the wafer and the total light detected. For example, in a proximity system the direction of the input light can be scanned with a galvanometer, causing the shadow of the

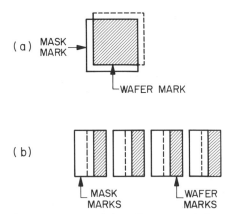

Fig. 13. (a) Mark for average transmission alignment technique. (b) Mark for average transmission alignment with improved signal.

window to traverse the wafer. In both cases the minimum is found either electrically (e.g., detecting the zero crossing of a differentiated video signal) or by computer fitting to the video signal.

(3) Further signal averaging can be obtained by periodic scanning of the image of the mark across the window by either of the above two techniques. With periodic, usually sinusoidal, scanning, three signal-processing techniques are available: (a) The total light integrated over even half-cycles is compared with the total light integrated over odd half-cycles. This method, sometimes called "boxcar" integration, improves the signal-to-noise ratio as the square root of the number of cycles over which the integration is performed. (b) The second harmonic of the scanning frequency is detected. This harmonic is present at alignment, since displacement in either direction changes the light intensity symmetrically. Maximizing the second harmonic is a sensitive test of the symmetry of the alignment. (c) The fundamental of the scanning frequency may be minimized.

Due to processing variations, the widths of the mark and the window may differ. Under these conditions, small relative motions produce no change in the total detected light. Although the scan must be large enough to cover this "dead band" in order to obtain a useful signal, there are no serious consequences on the alignment accuracy.

An obvious way to extend the sensitivity is to use multiple marks and windows (Fig. 13b); an array of narrow marks and slits offers more edges and hence more signal per unit area of alignment site. A periodic structure of equal dark and clear areas is called a Ronchi grating. While such a grating dramatically improves the signal level, it also reduces the capture range: If the alignment error is equal to the period of the grating, virtually the same signal is obtained as at alignment. Combinations of gratings of different periods have been used to retain most of the improvement in sensitivity while removing the ambiguity of a one-period error.

The advantages of these techniques are that signals are obtained in a format that is readily processed by analog or digital electronics and that high resolution is not required in the collection of the signal light. The principal disadvantage applies to those embodiments in which either the mask or wafer is physically scanned, especially if it is vibrated: The engineering used to make these motions must be of unusually high quality if they are to be precisely controlled and if the alignment system is to be mechanically reliable.

2. Scanning Microdensitometer

In a typical arrangement, a dark mark on the wafer is again viewed through a window on the mask. However, the window is now substantially wider than the mark (Fig. 11a), and both mark and window are in sharp focus on the detector. (In a proximity printer this may require the use of a bifocal lens; see Section III.C.) A video signal is obtained by scanning the combined image of mark and wafer across the detector.

At alignment the video signal is symmetric. A wide variety of detectors are suitable for this application:

(1) *A line detector* (e.g., photomultiplier or silicon diode viewed through a narrow slit). Scanning is done by moving the slit, by moving the entire detector assembly, or by moving the image of the marks across the slit with a galvanometer or other device in the imaging system (Fig. 14).

(2) *Line array.* This is a self-scanned array of up to several thousand photodiodes. In most arrays each photodiode is square; however, some arrays are available with very tall elements so that light can be collected from a wide swath through the mark.

(3) *Television camera.* Although at first this may appear to offer rather more information than is needed, the ability to average over multiple scan lines is an advantage.

Among these detectors, the photomultiplier–slit combination is the most sensitive (although comparable low-light television cameras based on multichannel plates or other image intensifiers are becoming available). The line arrays and television camera require the least mechanical engineering to implement properly, since they have no moving parts, with the galvanometer scan a very close second. The mechanical scan system and the line array have the best linearity, which is important for accurate interpretation of the video signal. The linearity of these scans may be limited by distortions in the microscope objective or other viewing optics,

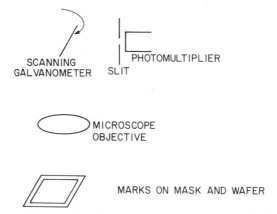

Fig. 14. Scanning the image of an alignment mark across a photomultiplier–slit detector.

perhaps at a level as high as 1 part per thousand. Apart from solid-state array cameras, the linearity of the television camera is likely to be the most difficult to control, and that of the galvanometer scan next most difficult.

If the mask and wafer are parallel, there are three degrees of freedom with which the mask can be aligned to the wafer (four, if separation, which may influence focus or magnification correction, is included). Since each of the systems described here measures only one degree of freedom, three (or four) such systems are required to achieve alignment. However, by using marks and windows at 45° to the direction of the television camera scan (or, equivalently, a "diamond within a diamond" in place of a square within a square), the two-dimensional character of the television scan can be used to measure alignment error in two degrees of freedom (Fig. 15). It is clear from the figure that only at alignment are all the video pulses equal in width. Only two such alignment systems, diametrically located on each side of a wafer, are required for alignment.

Although the scanning microdensitometer requires a fair amount of hardware, its implementation is straightforward. In many embodiments it resembles manual alignment, and the ability to design alignment marks that can be used with it and at the same time are compatible with manual alignment is an advantage.

If the alignment patterns consist of an array of narrow wafer marks

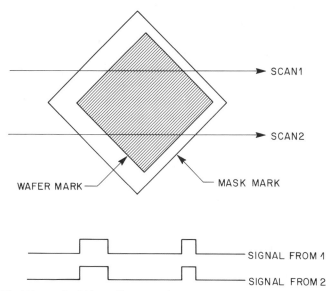

Fig. 15. Diamond within a diamond alignment mark. The presence of edges at ±45° permits alignment in two directions with only a single scan direction.

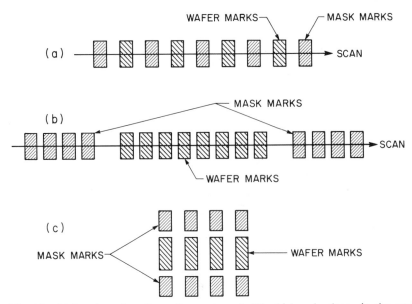

Fig. 16. (a) Interleaved gratings alignment mark. The high redundancy in the mark provides good signal averaging. (b) Adjacent gratings alignment mark for a linear detector. (c) Adjacent gratings alignment mark for a television detector.

centered within corresponding clear areas on the mask (Fig. 16a), the accuracy of alignment should be improved at the expense of additional processing. Note that in this periodic system one can detect when there is an error of one period, so there is not a gross reduction in capture range.

Alternatively, the wafer and mask arrays can be separated, permitting a more compact alignment pattern but in an arrangement more susceptible to optical distortion. Arrangements are shown that could be used with an accurate linear detector, for example, a line array (Fig. 16b) or with a television camera (Fig. 16c). In each case the phase of the two outside mask gratings is compared with the phase of the inside wafer grating, permitting a highly accurate determination of their relative position.

It should be noted that a distance $d = s^2/\lambda$, where s is the grating spacing and λ the wavelength of the (collimated) illumination, a grating forms an image of itself. The numbers are appropriate for use in a proximity x-ray printer, for example, to permit sharp focus simultaneously on the mask grating and a self-image of the wafer grating.

3. Setting the Gap

In some proximity printers, especially x-ray printers, the gap between mask and wafer is critical. This can be set mechanically or electrically (by measuring, e.g., the capacitance between the mask, or test patterns on the mask, and the wafer). However, purely optical methods are also used:

(1) A microscope objective is focused, either manually or automatically, on first the mask and then the wafer, and the distance between the two foci adjusted.

(2) A "light-section" microscope is used to project a narrow bright line on both mask and wafer along a direction that is 45° from the vertical. The two lines are then viewed at the complementary 45° angle, and the apparent separation between them is adjusted. A limitation here is that, because of the depth of focus required by the projection optics, the line will not be very sharp [Eq. (5)].

(3) A laser beam is focused to a point or line near the surface of the x-ray mask. Part of the light is reflected from the mask and subsequently diverges; another part goes through the mask, is reflected by the wafer, and diverges. Interference fringes may be observed in the region where the two diverging beams overlap. An accurate measure of the period of these fringes, using any of the scanning microdensitometer systems at a known distance from the mask, provides a direct measurement of the mask-to-wafer separation, which can then be set as desired [8].

4. Diffractive Alignment

a. Contrast Enhancement. The use of diffraction effects to improve contrast has already been discussed. By restricting the viewing optics so that only diffracted light is detected, the signal-to-background ratio can often be dramatically improved (Fig. 10). Analogs of all the conventional automatic alignment methods described above exist:

(1) *Average transmission.* The rectangular mark on the wafer is replaced by a grating. The grating is viewed through a window on the mask (for proximity printing) or imaged onto a window on the mask (for projection printing). In either case, spatial filtering in the detection or imaging optics restricts the light to that which has been diffracted by the grating on the wafer. Processing is as before, but with increased contrast and hence increased sensitivity.

(2) *Scanning microdensitometer.* Again, the wafer mark is replaced by a grating. Since the wafer mark is often narrow to begin with, it is common to direct the diffraction perpendicular to the alignment direction; that is, the wafer mark becomes a dotted line that diffracts the light along the direction of the line. As before, a scanning system is used to determine if the wafer mark is centered within the mask window, with spatial filtering being used in the optical path to improve the contrast and sensitivity of the system.

In general the diffractive element is found on the wafer rather than on the mask. This is because, ordinarily, mask contrast is *a priori* better than wafer contrast and also because many structures on the wafer that arise in normal processing are primarily phase structures. Such phase structures may have very low contrast, but quite high diffraction efficiency. However, care must be taken to avoid particular thickness of phase structures, where the path difference is a multiple of λ, and the diffraction efficiency drops toward zero.

The improvement in contrast in these diffractive systems is obtained at the expense of total light loss, perhaps as much as a factor of 10 or 20. A more serious limitation arises because the accuracy of alignment is dependent on the edge acuity of the alignment mark. For cases in which the diffractive direction is perpendicular to the alignment direction, the relevant edge of the alignment mark is defined by the ends of the lines in the grating (Fig. 10). Maintaining good acuity here is more difficult than maintaining it along the long side of a solid mark. However, if symmetry of the line profile can be ensured, the degradation in edge acuity may not be important.

b. Three-Dimensional Interference. Consider a grating on a mask located directly over a grating of the same period on a wafer in a proximity printer (Fig. 17). If parallel, monochromatic light is incident normally on the arrangement, the amount of light diffracted into various orders is a complex function of the grating parameters and the gap between the mask

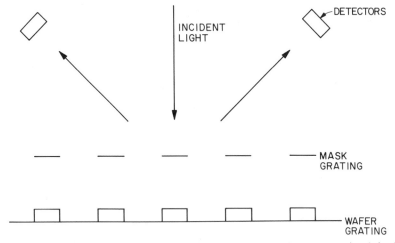

Fig. 17. Three-dimensional interference alignment mark. At alignment the signals in the two detectors are equal.

and wafer. However, by symmetry, it is obvious a priori that the first and negative first orders are of equal intensity. This equality is the basis of a diffractive alignment technique based on intensity measurements of the diffracted light [9]. A similar arrangement can be used in a projection printer, by imaging the wafer grating to the vicinity of the mask grating.

The technique is inherently sensitive, since the use of a small-period grating implies a rapid change of signal levels with misalignment. Such a small period might imply a capture range limited to half a period; however, the use of multiple gratings with different periods permits a relatively large capture range.

A more serious problem arises because the diffraction is a rapidly changing function of the mask-to-wafer gap. Although gaps exist at which the error signal is a smooth function of the misalignment, even small departures from these gaps produce significant degradation of the error signal. Another limitation is that any *blazing,* or nonsymmetric diffraction efficiency of the wafer grating, destroys the symmetry of the arrangement and introduces an error in the aligned position.

c. Zone Plate Alignment Systems. The properties of Fresnel zone plate patterns as diffraction-limited imaging devices have already been discussed. These properties arise because the total image-to-object path lengths through the different zones differ by integral multiples of the wavelength λ. It is easy to show geometrically that the boundaries between zones in a zone plate imaging a distant object are at distances r_n from the center of the zone plate given by

$$r_n = \sqrt{n\lambda f}, \qquad (10)$$

where f is the focal length of the zone plate and n is an integer. The resolution of the zone plate, approximately $\lambda \times f /$ number, is approximately equal to the width of the thinest, outside zone:

$$\lambda \times f / \text{number} \approx r_n - r_{n-1}. \qquad (11)$$

For the same f/number, as the focal length increases, the number of zones and the light-gathering power increase, but the resolution and the width of the thinnest zone remain the same. At short focal lengths the number of zones decreases; below 5 to 10 zones the imaging starts to degrade.

d. Linear Zone Plates. Linear zone plates are the analogs of cylindrical lenses; the ability to bring incident parallel light to a line focus permits their use in several methods of automatic alignment:

(1) *Position sensing.* A linear zone plate on a wafer in a projection printer can be used to form a line focus, which is relayed by the main

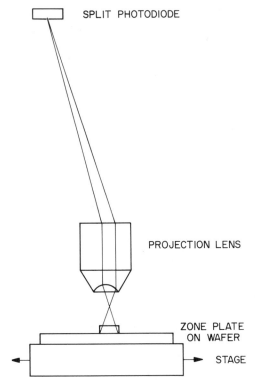

SPLIT PHOTODIODE

PROJECTION LENS

ZONE PLATE
ON WAFER

STAGE

Fig. 18. Zone plate alignment mark used to position chips on a wafer with respect to the projection lens.

projection lens onto a split photodiode (Fig. 18). The wafer position is then adjusted until a balance signal is obtained from the split photodiode, aligning the wafer with respect to the body of the projection printer along one axis.

(2)   The line focus formed by linear zone plates on the wafer and mask in a proximity printer can be viewed by any of the systems discussed under scanning microdensitometers, permitting an accurate measurement of the relative mask and wafer position (Fig. 14). The focal lengths are chosen to obtain simultaneous focus of the mask and wafer line images.

(3)   A linear zone plate on either the mask (Fig. 19a) or wafer (Fig. 19b) can be combined with a conventional mark on the other member [10]. If the line image and the mark overlap, the combination is viewed for total light transmission; if they butt, the combination is viewed with a scanning microdensitometer. In a proximity system the focal length of the zone plate is equal to the mask-to-wafer gap, which is somewhat short for comfortable

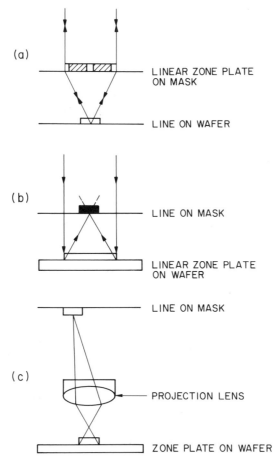

Fig. 19. (a) Use of a linear zone plate on the mask in a proximity printer. (b) Use of a linear zone plate on the wafer in a proximity printer. (c) Use of a linear zone plate on the wafer in a projection printer.

patterning of the linear zone plate. In a projection system (Fig. 19c) the focal plane of the zone plate must coincide with the mask or wafer plane. This can be achieved readily in refractive lens systems because of the substantial shift in focus of the projection lens between the projection wavelength and the (assumed much longer) alignment wavelength.

Much of the optics, particularly in the illumination systems, has been omitted from Figs. 18 and 19 for clarity. The arrangement in Fig. 19a can be used without any lenses; the linear zone plate serves us both a focusing and a collecting lens. A microscope objective is useful in Fig. 19b to collect whatever light passes the line on the mask. The parallel light used for illumination of the alignment patterns is readily obtained in systems with lenses by bringing a laser beam to a focused waist in the back focal plane of the lens.

e. Circular Zone Plates. Circular zone plates are the analogs of spherical lenses, bringing parallel monochromatic light to a diffraction-limited point focus. Like linear zone plates, they lend themselves to a variety of automatic alignment techniques:

(1) In position sensing, a circular zone plate on a wafer in a projection printer is used to form a point focus, which is relayed by the projection lens onto a *quadrant* detector, or photodiode split into four sectors (Fig. 18). The wafer position is then adjusted in both X and Y coordinates until all four sectors are balanced, aligning the wafer with respect to the projection printer in two directions. In this application it is important to keep the capture range small; that is, the quadrant detector should be nearly filled by the image of the zone plate spot. This serves to minimize contributions to the signal from light surrounding the focused spot; such light may not be symmetric and may cause biases in the aligned position.

Various devices exist in which an analog voltage indicates the position of an incident spot of light: Photodiodes and multichannel plate assemblies are available in which the position of the spot can be determined by how the output current divides between several output electrodes. The advantage of using these devices is that the zone plate position can be read without the wafer being moved, facilitating relative alignment between a mask and a wafer. However, the contribution of light surrounding the focused spot may be significant in these devices, and they must be used with care.

(2) With some restrictions, the scanning microdensitometer alignment approach is directly applicable to systems in which a circular zone plate is present on both mask and wafer (Fig. 14). In a proximity system the focal lengths of the two zone plates are chosen to differ by the distance between mask and wafer, so that the spots can be viewed in a plane in which they are both in focus. Similarly, in a projection printer the focal lengths are chosen to provide simultaneous focus.

One advantage of using a focused spot rather than a focused line from a linear zone plate is that alignment may be done in two directions with a single zone plate. The restriction is that a small circular spot is not an ideal match to some scanning systems. For example, a conventional line array will have a very narrow capture range (although arrays of wide photodiodes are available). On the other hand, scanning the focused spots across a slit provides an excellent signal with a reasonable capture range. There is a trade-off: The longer the slit is, the wider the capture range and the more degradation in the signal due to acceptance of background light.

Perhaps the most direct method of performing a two-dimensional scan of a zone plate spot is to image it on a television camera. The video signal can then be decomposed into picture elements small compared with the spot size and the centroid of the spot determined by computer processing. A major advantage of this method is that extraneous light surrounding the spot is entirely eliminated, so that an accurate result can be obtained. Disadvantages are the need for two-dimensional computer processing and biases introduced by nonlinearity in the television camera.

Solid-state television cameras, based on two-dimensional arrays of photodiodes, effectively eliminate distortion in the television camera (although not in the viewing optics), at some loss of sensitivity. Alternatively, the images can be combined by the optical system so they fall near each other, making the television camera distortion unimportant. Imaging *two* zone plates on the mask, equally and oppositely spaced from a single zone plate on the wafer, also helps to reduce the effects of distortion in the optics and camera.

(3) Combining a circular zone plate with a conventional mark is questionable; the small size of the matching conventional mark would make the system very susceptible to patterning errors.

Zone plate alignment in x-ray proximity systems enjoys a particular advantage: As shown in Fig. 8 x rays diverging from a point x-ray source produce a magnified image of the mask on the wafer, with the magnification a function of the gap between mask and wafer. In many alignment systems this gap must be tightly conrolled to prevent "runout" or magnification error. However, if the zone plates are illuminated at an angle such that the chief ray of the image they form is directed toward the x-ray source (Fig. 20a), then full alignment (in X and Y at two sites on the wafer) will automatically include the correct gap (Fig. 20b). The accuracy with which the gap is determined is such that the error in magnification is equal to the other alignment errors. An advantage of this procedure is that it automatically corrects for other sources of magnification error, such as mask or wafer distortion [11].

Zone plate alignment systems also suffer from a particular disadvantage: If the zone plate substrate is tilted or if the zone plate is covered by a wedged optical medium, the position of the focused spot will be shifted. A typical source of tilt, for an x-ray mask, might be "tenting" due to a nearby large particle caught between mask and wafer; a typical source of wedge on a wafer might be uneven resist flow. If the tilt angle of the zone plate is θ,

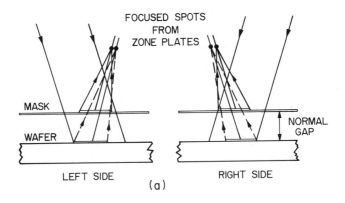

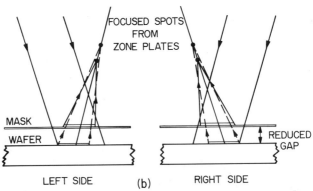

Fig. 20. (a) Circular zone plates on mask and wafer in an x-ray proximity printer. The gap is too large, resulting in a magnification error. (b) Same as (a), except that the gap has been reduced. The focused zone plate spots now overlap.

and its focal length is f, then the offset is given by

$$\text{Offset due to tilt} = 2f\theta \tag{12}$$

in the limit of small θ. If instead the zone plate is covered by material of index of refraction n, wedged at an angle ϕ, the offset is given by

$$\text{Offset due to wedge} = 2(n-1)f\phi \tag{13}$$

in the limit of small ϕ. The offsets may be significant. For example, if $\theta = 10^{-4}$ (a tilt of 1 μm/cm), then the focal position of a 250-μm focal length zone plate will be shifted by 0.05 μm. With vertical illumination, the tilt or wedge can be measured, and compensated for, by observing both real and virtual zone plate images, since they shift in opposite directions.

IV. MEASUREMENT OF ALIGNED WAFERS

A. Vernier Patterns

Vernier patterns (Fig. 12d), each half printed on a different lithographic level, are a time-honored method of measuring the overlay accuracy that was achieved between the two levels. Similarly, other patterns used for manual alignment (Figs. 11c,d) and techniques such as the use of Moire fringes are suitable for measuring the overlay accuracy after printing. The advantage of using these patterns is that only a microscope is required to make the measurement. The disadvantages are as follows:

(1) Measurement time is long, since an operator must make a judgment at each site measured.

(2) Measurements are limited in accuracy both by the definition of the patterns and by the address structure with which the patterns were written. The accuracy is often comparable to that which would be obtained in a manual alignment using the same patterns and is therefore of marginal usefulness.

B. Scanning Microdensitometer

A variety of scanning microdensitometers are available commercially, based, for example, on scanning slits or television cameras coupled to high-quality microscopes. Many of them were originally designed to determine feature linewidth by precise measurement of the video signal as the feature was traversed. Many of these instruments are ideally suited to measuring overlay accuracy between two levels, using a single mark on the first level centered between two similar marks on the second level.

Compared with their original use, in measuring overlay accuracy these instruments suffer a disadvantage: The marks being viewed are often in different focal planes, especially if a thick multilevel resist has been used [12]. This is probably more than compensated for by the fact that only the centroid of each mark need be located accurately, which is an easier task than finding the widths, as the machines were designed to do. Scanning densitometers are available with a number of convenient features such as automatic focusing and programmable stages, which facilitate the collection of a large amount of registration data in a short time.

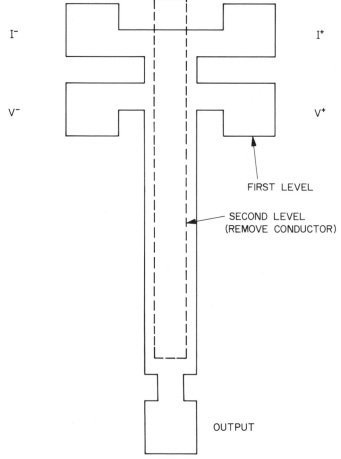

Fig. 21. Van der Pauw pattern to measure registration in one direction by removing conductor with the second level.

C. Electrically Probeable Patterns

Very high accuracy registration data can be obtained between two conducting levels. This is done by using van der Pauw patterns, which define an electrical bridge (Figs. 21 and 22) whose balance point is determined by the registration of the two levels [13,14]. The bridge terminals are brought out to pads, which are conveniently and rapidly probed by commercial equipment. If, as shown, separate current and voltage leads are used, the measurement accuracy is limited only by variations in the linewidth and in

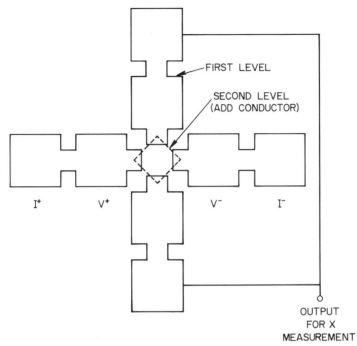

Fig. 22. Van der Pauw pattern to measure registration in two directions by adding conductor with the second level.

the sheet resistivity across the pattern, both of which may be determined with other probed measurements. The high accuracy, as well as the rapid data rate, make probeable patterns an ideal method of determining registration. The only significant drawback is that the method is limited to conducting patterns.

V. SUMMARY

This chapter describes a wide range of optical alignment techniques. The more complex marks have an inherent advantage for automatic systems: Because of the redundancy present in multiple edges, roughness in the pattern or local defects tend to average, and an alignment is obtained that is more robust to pattern degradation. This is particularly true of zone plate patterns, which can have major portions obscured with little degradation in their imaging. In addition, zone plates can form images from low-contrast phase objects, which are difficult to align otherwise.

Nevertheless, one cannot say that a particular mark or alignment strategy is superior. Alignment is just one part of wafer fabrication, and the entire process must be optimized. Good processing control and good engineering in the implementation of the alignment system are far more important than the choice of a particular strategy.

REFERENCES

1. F. A. Jenkins and H. E. White, "Fundamentals of Optics." McGraw-Hill, New York, 1957.
2. M. C. King, in "VLSI Electronics: Microstructure Science", (N. G. Einspruch, ed.), Vol. 1, pp. 41–81. Academic Press, New York, 1981.
3. A. Zacharias, *IEEE Trans. Components, Hybrids, Manuf. Technol.* **CHMT-5** (1982).
4. J. D. Cuthbert, *IEEE Solid State Technol. Workshop Scaling Microlithogr., New York, 1980; Microcircuit Eng., Aachen* pp. 190–197 (1979).
5. J. S. Courtney-Pratt and R. L. Gregory, *Appl. Opt.* **12**, 2509 (1973).
6. A. D. White, *Appl. Opt.* **16**, 549 (1977).
7. M. C. King and D. H. Berry, *Appl. Opt.* **11**, 2455 (1972).
8. D. C. Flanders and T. M. Lyszczarz, *Workshop Micrometer Submicrometer Lithogr., Indian Wells, Calif., 1983.*
9. S. Austin, H. I. Smith, and D. C. Flanders, *J. Vac. Sci. Technol.* **15**, 984 (1978).
10. B. Fay, J. Trotel, and A. Frichet, *J. Vac. Sci. Technol.* **16**, 1954 (1979).
11. M. Feldman, A. D. White, and D. L. White, *Opt. Eng.* **22**, 203 (1983).
12. J. M. Moran and D. Maydan, *Bell Syst. Tech. J.* **58**, 1027 (1979).
13. L. J. van der Pauw, *Philips Res. Rep.* **13**, 1 (1958).
14. D. S. Perloff, *Solid State Electron.* **21**, 1013 (1978).

Chapter 7

Metrology in Microlithography

D. Nyyssonen†

National Bureau of Standards
Gaithersburg, Maryland 20899

† Currently with CD Metrology Inc., Germantown, Maryland 20874.

I. INTRODUCTION

The advent of VLSI and VHSIC technologies has had an enormous impact on the requirements for accuracy and precision in dimensional metrology. Accurate and precise measurements are needed to improve yield, to ensure that lithographic and critical dimension (CD) measurement systems meet specifications, and to establish control of fabrication processes. The push to micrometer and submicrometer feature sizes on larger and larger wafers, now approaching 250 mm (10 in.), demands routine distance measurements for overlay and registration with accuracy approaching one part in 10^8 as well as measurements of features as small as 0.5 μm. Dimensional measurement systems require precision and accuracy that can take up only a small fraction of the error budget of 10% or less allowed in process control. By the gauge maker's rule, the measurement system tolerance must be 3–10 times less than the tolerance on the part being manufactured. To meet these goals, CD measurement systems must be critically evaluated.

This chapter discusses primarily distance measurements both (1) across a wafer, photomask, or reticle, such as required for overlay or registration and evaluation of lithographic systems, and (2) feature-size measurements, typically for dimensions less than 10 μm. Each of these measurements can be made in one of two ways, each with distinctively different sources of metrological errors: (1) on an image or in the image plane of a microscope and (2) in the plane of the wafer, as in the case of interferometric measurement of stage position. In both cases, two coordinate systems can be established, one in the *actual* plane where the measurement is made and the other in the plane of the *desired* measurement, that is, the object plane or plane of the wafer. In an ideal measurement system, these two coordinate systems are related very simply, mapping onto one another linearly. For example, in a feature-size measurement using an optical microscope, one of the coordinate systems is located in the focal plane of the microscope and is assumed to be coincident with the feature on the wafer being measured. The other is located in the image plane, and the two are assumed to be related only by magnification and some zero offset.

In fact, all types of distortions and nonlinear relations of these coordinate systems may occur because of the mechanical, optical, and other imperfections of the measurement system, which inevitably limit the accuracy and precision of the measurement. In addition, feature-size measurement has additional sources of error associated with the method of edge detection. This chapter attempts to provide a framework for discussing the sources of error in these types of dimensional measurements and for assessing the current capabilities and needs of integrated circuit (IC) metrology tools.

II. ACCURACY AND PRECISION

Before dimensional measurements are analyzed, a discussion of accuracy and precision is needed. In conventional metrology [1], precision, or repeatability, is defined as the spread in values associated with repeat measurements on a given sample. That is, the measurement of a given quantity will produce measurements that can be averaged to produce a mean value

$$\bar{x} = \sum_{i=1}^{n} x_i/n, \tag{1}$$

where x_i is the result of the ith measurement and n the total number of measurements. The precision, or repeatability, is characterized by the standard deviation

$$s = \left[\frac{\sum_{i=1}^{n} (x_i - \bar{x})^2}{n - 1} \right]^{1/2}. \tag{2}$$

These general formulas assume that n is large and that the errors are random and result in a Gaussian or "normal" distribution centered about the mean, as shown in Fig. 1a. In many cases, such as length metrology, this may not be true. For example, a common source of error in dimensional measurements is misalignment of the target to be measured to the axis of the measuring instrument. In this case, misalignment always causes measurements that are too large, so that, if misalignment were the only source of error, the distribution would be one-sided, as shown in Fig. 1b. Fortunately, misalignment is only one source of error, and the resulting distribution from all sources of error may still appear nearly normal if all other sources of error are randomly distributed, as in Fig. 1c. When the distribution is clearly not normal as shown by a histogram, care has to be taken in summing errors or correcting calibration data since the usual rules may not apply [1].

A. Short- versus Long-Term Precision and Accuracy

Of major concern is the interpretation of a given standard deviation or statement of precision for a measurement system. Does it guarantee repeat-

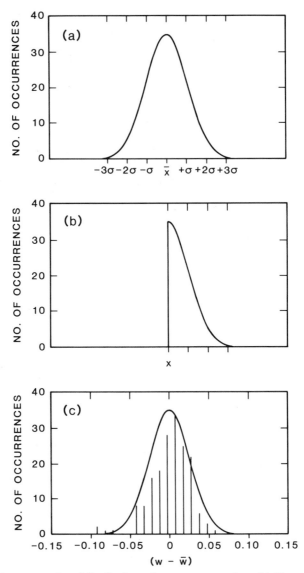

Fig. 1. Some examples of distributions of measurement values. (a) Normal distribution (all error sources are random); (b) one-sided distribution such as might occur in dimensional measurements where the major source of error was random misalignment; (c) a real nonsymmetric distribution (taken from control data on line-spacing measurements at 10.8 μm, courtesy of Marilyn J. Dodge, NBS) with normal distribution (best fit) shown for comparison. Discrete values result from digitization of measurement values.

ability from day to day or week to week? The answer depends on two considerations: (1) over what period of time the measurements were made to determine the quoted precision and (2) how often the system is rechecked or recalibrated. If the system is rechecked, readjusted, or recalibrated at least once every day, then precision determined by measurements taken over a time period of one day may indicate how reproducible measurements will be between rechecks or recalibrations. On the other hand, if the measurements on which precision were based were taken 5 min immediately after calibration of the system, the quoted precision is probably meaningless on a day-long time scale.

If the system is expected to hold calibration over periods of weeks or months without recalibration, perhaps taking only control chart measurements as would be done in a standards laboratory using the measurement system regularly, the only meaningful precision statement must be based on measurements taken over a long period (i.e., weeks or months). For example, a system may be highly repeatable over a period of a few minutes but be extremely temperature dependent such that temperature fluctuations during the course of the day may produce much larger measurement variations. Therefore, measurements made to determine precision and accuracy must match the intended use of the system.

The difficulty with the longer time period for calculating precision is that long-term precision is generally poorer than the short-term precision quoted by the manufacturer and may reflect not only long-term instability or drift in the measurement system but also environmental effects, which may vary from site to site. In dimensional measurement systems, more attention must be paid to the environmental control needed at the user's site to maintain a given precision as quoted by the manufacturer.

Accuracy is a much less well defined concept than precision. Usually, there is some agreed-upon quantity that one is trying to measure. However, when examined in detail, this quantity and its definition frequently become fuzzy and may escape clear definition.

For example, linewidth on IC features seems like a clear enough idea until one begins to look at real structures. In Fig. 2a, the line has an ideal structure with vertical walls and smooth edges, and the linewidth can be unambiguously defined. Real structures, like that shown in Fig. 2b, do not have well-defined edges. They may have asymmetric, nonvertical geometry with raggedness along their length. In different applications, the quantity that is to be measured may be different, for example, the width at the bottom when either etching or doping will be the next process step or the mean width if comparisons with electrical linewidth measurements will be made. Moreover, the only meaningful measurement may be an average

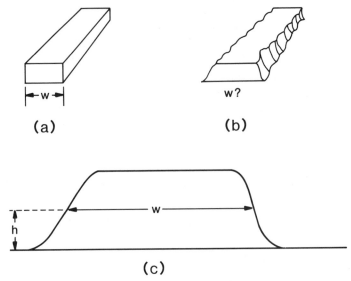

Fig. 2. Definition of linewidth. (a) Ideal line geometry with width W unambiguously defined; (b) real line structure showing asymmetric, nonvertical edge geometry and edge raggedness; (c) proposed definition of linewidth as width W defined at some height h above the interface (between the patterned layer and sublayer) and averaged along some length of the line. The height h selected is appropriate to the application (e.g., near the interface for patterned resist).

along some specified length of the line when grain structure or raggedness from uneven etching is present.

Therefore, a more refined definition of linewidth is needed. For instance, we may agree, as has been proposed, to measure the geometric width of the line structure at some distance above the interface, averaged along a 1-μm length, as illustrated in Fig. 2c. The problem then becomes one of determining how well a given instrument can measure the agreed-upon quantity. If the system measures such a quantity with a systematic error, that is, it always measures too large or too small, we can determine the average error or offset from measurements on a reference standard with known values. We define this average error as the accuracy of the measurement. Although this average error may be so small that we choose to ignore it, it probably is never zero in any real system. It is the job of the engineer or metrologist to assess its magnitude realistically.

In the case of many engineering standards, users may agree not on the quantity to be measured, but on a measurement technique, including the method, system configuration, and sometimes even the manufacturer and specific model number of the system to be used. In this case, the difficulties of defining what is being measured and assessing the accuracy are avoided. However, one drawback of this approach is that, if any changes are made in the system, as with newer models, or changes occur in the characteristics of the sample being measured, or systems vary from one to the next, the system may no longer be measuring the same quantity, and process control may be lost. Another drawback is that the measurement results may be difficult to relate to theory or use as input data to modeling or simulation programs.

The ideas of accuracy and precision can be combined into what we call here the total uncertainty (see Fig. 3) of a given measurement [2]:

$$U = E + 3s. \tag{3}$$

The meaning of the total uncertainty, then, is that a given measurement cannot be stated to be any closer than U to the quantity that is to be measured (worst case) since there is a finite probability that it could differ by that amount. When a measurement is given as $x \pm U$, the desired quantity may lie anywhere in the interval defined by $\pm U$. It is not possible to "know" that quantity any better without additional measurements (to improve precision) or proper calibration to a known reference standard (to

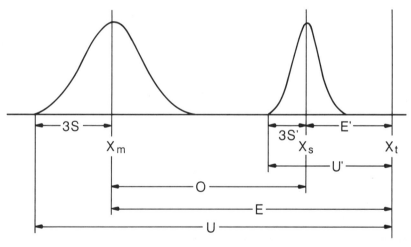

Fig. 3. Definition of uncertainty U and standard deviation s. In this figure X_t is the "true" value or desired value of the measurement, X_s the value assigned to the standard with its precision given by $3s'$ and total uncertainty U', and X_m the result of measurement on another system with precision $3s$. If the measurement offset O is eliminated by correction to the value of the standard X_s, the uncertainty U associated with X_m is still at least $U' + 3s$. Note that X_t is frequently ill-defined and that, when the characteristics of the standard used to determine E do not match those of the part to be measured, the uncertainty in X_m may actually be larger than indicated.

reduce the systematic error in the measurement). It must be clearly understood that when we cannot assess the accuracy of a given measurement
system as illustrated in Fig. 3, all we can talk about is precision, and we
cannot meaningfully compare measurements made by different systems or
different techniques.

If more measurements are made, we can average them and improve the
precision in Eq. (2) by dividing by $\sqrt{n}$. The accuracy E will remain the same
unless we can provide a correction to the data by calibration to a reference
standard. Again, even when a reference standard is used, E cannot be
reduced to zero. The calibration standard has some stated accuracy and
precision associated with it (Fig. 3):

$$U' = E' + 3s'. \tag{4}$$

When measurements are corrected by subtracting the average difference
between the known values and results of measurements on the reference
standard, the total uncertainty becomes

$$U = E' + 3\sqrt{s^2 + s'^2} \tag{5}$$

The rule here is that random errors are added in quadrature, but systematic
errors must be added linearly [2].

Even when measurements are not corrected to the reference standard
because the measured values lie within the stated uncertainty of the standard, the measurement system cannot be stated to have an uncertainty less
than the standard to which it is compared. For example, in linewidth
measurements such as those claiming traceability to the National Bureau
of Standard's (NBS) 474 or 475 photomask standard [3], the total uncertainty ± 0.05 μm assigned to these linewidth standards arises not from any
significant error in the NBS calibration system, but almost completely
from the ambiguity of the definition of linewidth when the edge geometry
deviates from vertical. In normal fabrication of these standards, as with all
photomasks, the edge geometry or slope of the line pattern varies somewhat. To improve the uncertainty associated with the standard, a more
precise definition of linewidth must first be adopted. Tightening the specifications on the edge geometry of the photomask reference standards
would not solve the problem as long as photomasks produced in the
industry had varied edge geometries and the variations were not properly
taken into account in the linewidth measurement.

B. Total Uncertainty versus Precision

Many engineers argue that, in process control measurements, all that is
needed is repeatability, or precision, in a measurement system. That is,

they believe that as long as process control measurements are repeatable, absolute accuracy does not matter. Although in many cases this may be true, in other cases this assumption may produce gross errors, great difficulty in controlling the process, and difficulty in relating process parameters to predicted device performance.

In particular, when a given measurement is influenced by properties of the part being measured or these properties do not match those of the reference standard, variations encountered in normal processing may cause the systematic error E to vary with variations in the part in ways unknown to the process control engineer. As a result, parts that are actually in error may appear from the measurement to be good, and, conversely, some that are good may produce measurements that indicate large errors. For example, optical linewidth measurements are generally affected by such variables as the geometric cross section of the line being measured, the thickness and index of refraction of the patterned layer and sublayers, and the granularity and uniformity of the materials. Thus, if a single reference standard is used to determine E and s, these values will not be representative of the true E and s encountered in measurements on a part that does not match the reference standard. The fabrication process may be out of control with unacceptable part dimensions going undetected due to measurement errors, even though the instrument appears to have excellent precision when measuring the reference standard.

Within this framework, we shall now look at some of the dimensional measurements used in both IC process control and evaluation of microlithography systems, as well as methods of determining sources of metrological errors.

III. IMAGING IN THE OPTICAL MICROSCOPE

Because the microscope in one form or another is an indigenous part of optical instruments used for measuring IC photomasks and wafers, some discussion of its imaging characteristics with regard to metrology is necessary. There are many excellent books and articles [4–6] covering many aspects of optical imaging as applied to microscopy, but these rarely discuss the special requirements of dimensional metrology. We include here a discussion of only those aspects of imaging in the microscope that are important for metrology and that differ from conventional requirements. Although the following discussion may seem extremely detailed, it must be remembered that IC fabrication tolerances of 10% or less on micrometer and submicrometer features require elimination of sources of metrology errors of 0.01 μm (100 Å) or less.

Today's optical measuring microscope evolved from a strictly visual and later a picture-taking instrument that allows us to look into the microworld in ways that would not be possible with the unaided eye. However, to achieve accurate metrology at micrometer dimensions, we have to stop thinking of microscopes as picture-taking instruments and look at them as waveform analyzers or image analyzers. Exactly what waveforms are produced for given features patterned in a given set of materials is extremely important. Exactly how these waveforms are produced and how they are analyzed is the key; image enhancement (by which we mean contrast enhancement, differentiation, filtering, etc.) and sharp, "nice" pictures are antithetical to accurate metrology.

The requirements placed on the imaging system and waveforms it produces are quite different depending on whether we must (1) simply detect the presence of a given feature, as in defect inspection; (2) measure the distance between two similar features, as in overlay or line-spacing measurements; or (3) measure the size of a feature, as in linewidth measurement. In the following discussion of the imaging system, we shall note when possible the differing requirements for these different types of measurements.

A. Coherence

Before we can discuss resolution and other aspects of imaging in the microscope, we shall discuss one of the fundamental aspects of imaging micro-objects: spatial coherence.

The coherence parameter s_B that has come to be used to describe the interference properties of conventional bright-field microscope imaging is defined as the ratio of the numerical apertures (NAs) of the condenser to that of the objective lens in the microscope imaging system,

$$S_B = NA_{cond}/NA_{obj}. \tag{6}$$

and is illustrated in Fig. 4. A fully coherent, bright-field imaging system has a very small S value ($S_B \rightarrow 0$), while an incoherent system has a very large value ($S_B \rightarrow \infty$). In conventional imaging in the microscope, neither one of these limits is ever realized. Instead, we talk about effective coherence and effective incoherence. For effective coherence,

$S_B \leq \frac{2}{3}$,   for high-contrast objects such as opaque photomasks,

$S_B \leq \frac{1}{5}$,   for low-contrast objects with phase variations (most wafer features).

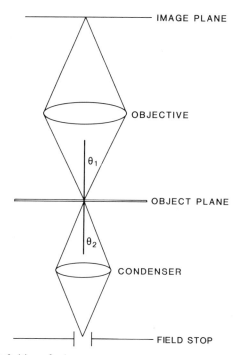

Fig. 4. Definition of coherence parameter,

$$S_{\mathrm{B}} = \mathrm{NA}_{\mathrm{cond}}/\mathrm{NA}_{\mathrm{obj}} = \sin \theta_2/\sin \theta_1 \qquad \text{(dry optics assumed)}.$$

For effective incoherence,

S_{B} may have any value for low-contrast objects with no phase variations present (rare),

$S_{\mathrm{B}} > 2$,   for all other cases.

For practical purposes, the images formed for these values of S_B would be indistinguishable from the corresponding completely coherent and incoherent images. One consequence of these considerations is that a high-NA conventional bright-field imaging system with a 0.9-NA dry microscope objective can *never* be effectively incoherent with $S_{\mathrm{B}} > 2$; we must therefore deal with partially coherent or effectively coherent imaging with such a microscope.

For focused-beam scanning systems with the roles of the illumination and detection systems interchanged, the coherence parameter is defined as

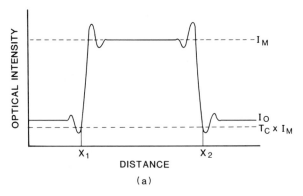

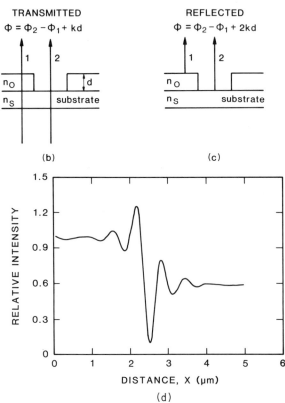

Fig. 5. Definition of parameters T (or R) and ϕ used to characterize line objects. (a) Image profile showing I_M and I_O, where T (or R) $= I_O/I_M$. When T_C times I_M is used for edge detection, the linewidth is given by $X_2 - X_1$. The optical phase difference ϕ is determined from the optical path difference between rays 1 and 2 for (b) transmitted light and (c) reflected light. In each case, the values of ϕ_1 and ϕ_2 can be calculated from the given complex indices of refraction n_o and n_s and the thickness d of the etched layer by using the Fresnel equations. (d) Coherent edge image for $T = 0.6$, $\phi = 0.6\pi$

the inverse of that given above, that is,

$$S_F = NA_{coll}/NA_{fb} \tag{7}$$

(Nyyssonen [7]). For such systems, effective coherence and incoherence are similarly defined for this S_F. Therefore, it is possible to produce a system with an effectively coherent or incoherent response using either a laser or a thermal source such as a tungsten–halogen lamp.

One major difference between coherent and incoherent imaging is sensitivity to phase changes in the object or feature being measured. In Fig. 5, we illustrate the parameters R (normalized reflectance) and ϕ (phase) used to describe the optical properties of the object. Incoherent imaging systems "see" only variations in the reflected intensity R, while coherent systems "see" variations in the phase as well. In many cases, where there is very little intensity contrast (corresponding to $R \rightarrow 1$), all of the signal is contained in the phase variations. In such cases, the dark interference fringe that occurs at the line edge as illustrated in Fig. 5d is the result of the phase discontinuity at the line edge as seen by the coherent imaging system.

B. Resolution

Over the years, many criteria have been developed for describing the limit of resolution of an optical system. Two commonly used ones are illustrated in Fig. 6. Unfortunately, as has already been realized for lithograph exposure systems, neither of these criteria is particularly useful in describing the limit of resolution of a dimensional measurement system. The requirements are different for line-spacing or pitch measurements as compared with linewidth. The "resolution" required for distance or line-spacing measurements where the line image is used merely as a fiducial mark is much less strigent than that required for feature-size measurement. In linewidth measurement, accurate edge detection requires an imaging system that introduces very little loss in modulation or contrast.

For line-spacing or distance measurements, if the lines are well separated (by a few micrometers or more), it is important only that the line images be identical. In fact, linewidths of 0.1 μm, well below the conventional Rayleigh resolution limit, in thick layers such as 1-μm resist, can be imaged and the distance between them measured optically. A line-spacing measurement is also tolerant of all sorts of poor image characteristics — that is, the line images may be out of focus, they may differ in shape from day to day, they may be noisy and the image smoothed, or they may be subject to vibration from the environment — as long as the images are identical and the center or other reference point on one line (as defined by the measure-

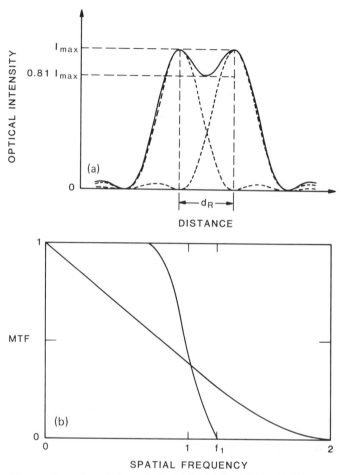

Fig. 6. Commonly used resolution criteria. (a) Rayleigh resolution limit corresponding to separation d_R of two unresolved bright points (incoherent with respect to one another) when two distinct points are discernible by the optical system; (b) resolution limit defined as a spatial frequency at which the modulation transfer function (MTF) of the imaging system falls to a value of zero; $d = f \lambda / NA$. The spatial frequency f_1 is the resolution limit for an effectively coherent system; $f = 2$ corresponds to that of an incoherent system.

ment system) remains fixed with respect to the other line. On the other hand, the accuracy of feature size measurements depends critically on the image structure and the use of an accurate edge detection criterion. This is one of the reasons that accurate and reproducible magnification calibration does not guarantee accurate or reproducible linewidth measurements in either an optical microscope or a scanning electron microscope (SEM).

For distance measurements between isolated lines, then, the resolution limit (ability to detect a narrow line) depends solely on the ability to detect a measurable and reproducible signal, and the width of the lines may be well below $\lambda/10$ for a very sensitive photodetector, where λ is the illumination wavelength.

As the narrow lines become closer together, the accurate measurement of their separation distance is impeded by other difficulties due to proximity (interaction) effects [8]. The center of the line image varies with respect to the actual line center (Fig. 7). This shift of the image away from the true line center is greatest for coherent imaging. Depending on the accuracy required, the useful resolution (center-to-center line separation) may be limited to λ/NA for coherent imaging ($0.6\lambda/NA$ for incoherent imaging), well below both the modulation transfer function cutoff (Fig. 6b) and the Rayleigh resolution limit (Fig. 6a). Since this shift usually affects only the outermost lines in a multiline target, the problem can be eliminated by including dummy lines (not used for measurement) in the design of the target.

For linewidth or feature-size measurement, the requirements are quite different than for pitch or line spacing. In an effectively coherent, bright-field imaging system, linewidths W can be accurately measured for

$$W \geq 0.85\lambda/NA. \qquad (8)$$

(Adjacent lines must be separated by at least the same limit W for accurate measurements [9].) In some cases, if assumptions can be made about the

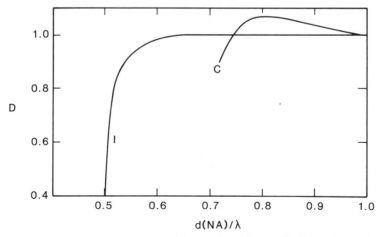

Fig. 7. Ratio D of the measured-to-real line spacing D for an effectively coherent imaging system (C) and an incoherent system (I).

nature of the line object and the reproducibility of the line shape from one line to another, dimensions well below this limit can be measured by special techniques [10]. Note that, when a line- and spacewidth are added to produce a pitch or line-spacing measurement, the tighter resolution and other requirements for feature-size measurement must be followed to ensure that they sum to the correct pitch.

Confocal microscopy [6] has also been proposed as a technique for linewidth measurement. It is a coherent imaging system that has the following advantage. Its impulse response is the square of that of a normal bright-field microscope, which has the effect of reducing the contrast of the bright fringes near the line edge without affecting the dark fringe or minimum at the line edge. Although a confocal microscope has increased resolution for low NAs where $\tan \theta \simeq \sin \theta$, the improvement in resolution is negligible for the high NAs used for linewidth measurement [7].

C. Kohler Illumination

In addition to detectability and resolution, linewidth metrology requires certain image characteristics; one of the most important is symmetry in the image. Because coherent imaging systems are sensitive to phase, it is important that the illumination not only be uniform in intensity, but also uniform in phase at the object plane of the microscope. Because of this, accurate alignment of the optical system is required. One of the easiest ways to ensure that these requirements are met in a bright-field imaging system is to follow the procedures for setting up Kohler illumination [11]. As illustrated in Fig. 8, Kohler illumination requires that (1) all elements of the microscope, including apertures and lenses, be centered on the optical axis; (2) the field diaphragm be in focus at the object plane; (3) the aperture diaphragm be in focus at the back focal plane of the imaging lens; and (4) diaphragms be set properly to achieve the desired degree of coherence. In addition to ensuring uniform symmetric illumination with uniform phase,

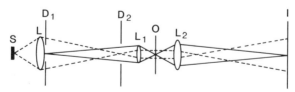

Fig. 8. Kohler illumination showing the imaging of the source S onto the back focal plane of the condenser lens L_1 and the imaging of the field diaphragm D_1 at the object plane O. Coherence is determined by adjustment of the aperture diaphragm D_2 relative to the aperture of the imaging lens L_2.

Kohler illumination provides the most energy-efficient illumination in a standard microscope.

D. Aberrations

Elimination of aberrations in the optical system is much more important for feature-size measurement than for distance measurement: Only asymmetric aberrations (such as coma) and distortion are of concern in linescale or distance measurements. Since all aberrations, including defocus, distort the optical image, nearly all feature-size measurements are sensitive to their presence. The best test is to compare the image profile in an actual system with that predicted for a diffraction-limited system. For a large amount of aberrations, the image has easily identifiable characteristics such as shown in Fig. 9a for spherical aberration. However, even when no gross image defect is discernible, resolution or ability to measure micrometer and submicrometer features may be degraded. The effect of limited resolution on linewidth measurement is shown in Fig. 9b.

Distortion is important in accurate line-spacing as well as feature-size measurements when made in the image plane of the microscope and is most readily measured with a two-dimensional grid or linear micrometer scale. Although a small amount of distortion may not be discernible to the eye, the effect is to change the measured distance or linewidth as the feature is placed in different positions within the field of view. In some cases, restricting the field of view to the region where the error is tolerable is sufficient. Measurements of distance made interferometrically by moving the wafer rather than scanning the image avoid errors due to off-axis aberrations. In the case of feature-size measurements, however, the symmetric aberrations, spherical and defocus, affect edge detection.

E. Focus

Given a good-quality optical system, properly aligned, the most significant remaining source of error in a feature-size measurement is inaccurate focus. Because of this, most CD-measuring instruments have incorporated some form of automatic focus. Sensitivity of the measurement system to defocus varies with both the edge detection criterion used and the NA of the imaging objective and degree of coherence. Coherent systems are more sensitive to defocus, especially when phase variations are present in the object (most wafer materials and resist). The conventional Rayleigh depth of focus [12] (developed for incoherent imaging) does not apply here. Focus tolerances have to be developed individually for a given edge detec-

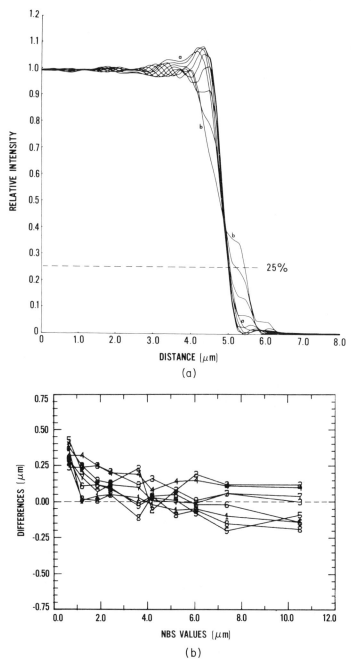

Fig. 9. (a) Theoretical edge image profiles with defocus and 2λ spherical aberration. Defocus varies from (a) 0 to (b) -2 waves in steps of $\lambda/4$ (0.22 condenser NA, 0.65 objective NA). (b) The effect of poor resolution on linewidth measurement; small lines measure too large.

tion criterion. For example, in the literature, there has been reference to a so-called isofocal point, or threshold in the image profile, which is relatively insensitive to defocus [13]; that is, linewidth does not change as rapidly with defocus when the isofocal point is used for edge detection. It has been shown [13] that the coherent edge detection threshold is such a point.

A wide variety of focus-sensing techniques are used in metrology systems ranging from tests on the image profile, such as steepest edge image slope, to surface-sensing techniques, including range finding and sensing of the intensity in the reflection of a focused beam from the surface. For calibration of photomasks, NBS has been using the dual-threshold focusing criterion shown in Fig. 10a [14]. For nearly opaque photomasks, the higher threshold is taken as 90% of maximum intensity, and the lower threshold at 10%. The distance Δ can be determined analytically for a diffraction-limited optical system with a given NA where the line has vertical edges. In addition, a tolerance can be placed on the acceptable range of values of Δ required to maintain a given accuracy and repeatability for linewidth measurement. This tolerance can be determined by correlation of Δ with linewidth since linewidth increases (or decreases) with defocus when a fixed threshold is used for edge detection. A low correlation coefficient indicates that measurement errors other than defocus dominate. Too high a correlation coefficient indicates that incorrect measurements are consistently being made on out-of-focus images. In such a case, linewidth precision and accuracy can be improved by tightening the focus tolerance or improving the sensitivity of the focusing mechanism.

In addition to defocus, Δ is influenced by the edge geometry of the mask. On a poor-quality mask, where etching has produced nonvertical edges with appreciable slope or raggedness, Δ will be larger than predicted for ideal vertical walls. Therefore, rejecting measurements for which Δ is larger than a calculated value places a specification on mask quality as well as accuracy of focus.

Some data accumulated on NBS photomask standard reference materials (SRMs) are shown in Fig. 11. The variation in linewidth (measured width minus the mean width) is plotted as a function of Δ for a large number of measurements on the same photomask. The data shown exhibited no correlation between focus and linewidth indicating that variations in linewidth are attributable to sources of error other than defocus.

A similar dual-threshold technique has been applied to patterned thin layers on wafers [14], as illustrated in Fig. 10b. However, for thick layers such as resist and most metal features on wafers, focus determination is complicated by "waveguide" effects [13,15]. The anomalies that result (e.g., no change in steepest image edge slope with large changes in width)

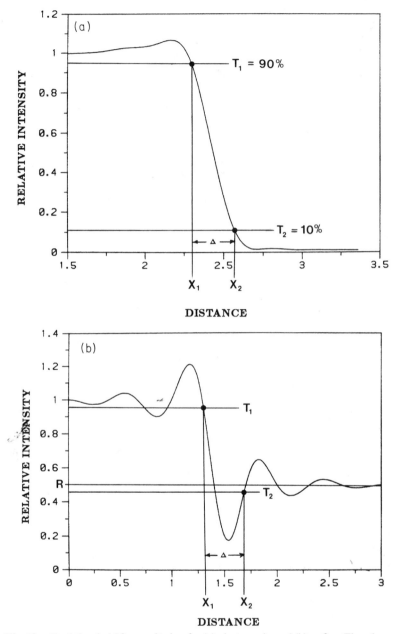

Fig. 10. Dual-threshold focus criterion for (a) photomasks and (b) wafers. The edge width Δ is a minimum at focus. For wafers, $T_1 = 0.95 I_M$ and $T_2 = 0.95 I_O$ (see Fig. 5).

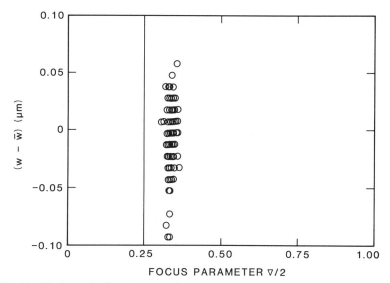

Fig. 11. Variation in linewidth as a function of the focus parameter Δ for a group of photomasks calibrated on the NBS system described in Nyyssonen [16]. The data showed no significant correlation between W and Δ, indicating that focus is not the dominant source of error in the measurements.

render criteria based on image structure insensitive. At this time, no single acceptable focus criterion has been developed that is applicable to all wafer materials and geometries.

F. Flare

Some flare or stray light is unavoidable in all optical systems. By flare, we mean here any non-image-forming light that adds an incoherent background to the image. In a good system, it may be 1% or less of the image intensity, while in a poor system it may be 10% or more. The amount of flare is also dependent on the reflectivity of the mask or wafer materials. Figure 12 shows an example of a bright chromium photomask (Fig. 12b) measured in a system that has been calibrated with an antireflective chromium mask for which no evidence of flare was apparent (Fig. 12a). In this case, the amount of flare can be estimated from comparison of the relative reflectances of the mask and substrate materials in a line image versus a same-size window image, as shown in Fig. 13. If the amount of flare is known, it is possible in some cases to correct the ensuing mask linewidth measurements. However, it is preferable, especially in video-based systems,

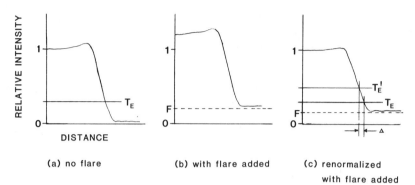

Fig. 12. Effect of flare F and the resulting edge detection error Δ when no correction is made there. T_E is the threshold used for edge detection when no flare is present and is incorrect is case (c). T_E' is the correct threshold when flare is present.

to calibrate the system using a mask of the same material (thickness and refractive index) as that for which measurements are to be made [16].

G. Spectral Bandwidth

The requirements on the spectral bandwidth of the optical system, including illumination and detector, vary with the materials being measured, the thickness of the patterned layer, substrate materials, and the type of measurement. Spectral bandwidth is of little concern in a line-spacing measurement (unless chromatic aberrations produce an asymmetric image), whereas, as with other optical parameters, it is extremely important for feature-size measurements, especially on thick patterned layers such as resist. For linewidth measurements on thin, spectrally neutral materials, the major concern is image quality. Optical microscope objectives are typically designed for visual viewing and have their best response at a wavelength of ~ 530 nm. Hence, a green filter that peaks the instrument spectral response (including light source and detector) at this wavelength is generally recommended, unless, of course, the optics have been designed for or corrected at another wavelength.

In addition, spectrally nonneutral materials vary in appearance as quantified by the reflectance R and phase ϕ, which vary with wavelength. In general, the thicker the layer, the narrower the spectral bandwidth must be to maintain a sharp image and uniquely define R and ϕ. Also, the image structure may be sensitive to changes in thickness, as in resist, when a broader spectral bandwidth is used. Figure 14 shows the variation in R and ϕ for a 0.6-μm-thick silicon dioxide layer on silicon as a function of wavelength. Since R and ϕ relate to image structure (see Fig. 5d), the

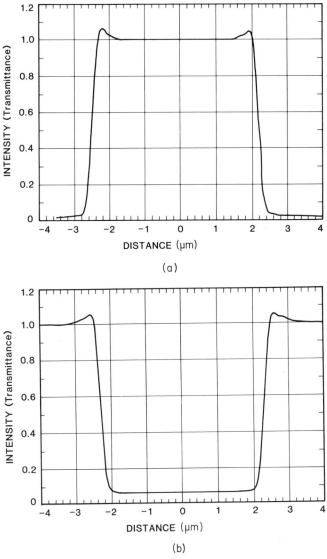

Fig. 13. Experimentally measured image profiles of (a) a window and (b) a line on a bright chromium photomask showing the difference flare makes on the value of I_0.

variation in R and ϕ with wavelength causes changes in image structure, which are effectively integrated over the spectral bandwidth of the system. Because of the spectral response of most detectors, including photomultipliers and video cameras, the illumination is effectively filtered to the equivalent of a broadband green filter. For this reason, the effects of the

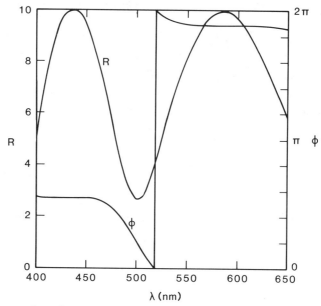

Fig. 14. Relative reflectance I_O/I_M (curve R) and phase differences ϕ (curve ϕ) for a 0.6-μm-thick layer of silicon dioxide on silicon calculated from the Fresnel equations for varying wavelength. Curves are normalized with respect to the parameters of the silicon substrate.

variation of R and ϕ with wavelength, although present, especially on thin layers (less than $\lambda/4$), may not be as readily apparent as the effects of coherence or angle of illumination.

H. The Detector

The spectral response of the detector has been discussed above. In addition, the linearity and slope of the detector versus intensity response curve and the response range are also important. Of the two basic types of detectors (video camera and slit–photomultiplier), the trade-off is between speed and convenience, on the one hand, and photometric accuracy, on the other.

Clearly, a nonlinear response curve distorts the image waveform. What is not so obvious is that a log slope (or γ) other than unity affects feature-size measurement. If an optical threshold is used for edge detection, a slope other than unity changes the effective threshold. In practice, a known linewidth from a reference standard can be used to adjust the edge detection threshold to yield the correct measurement. In this case, it is important only that the slope of the response curve, light level, and flare or stray light level be held constant when the unknown is being compared with the standard.

Since the optical threshold used for coherent edge detection [13] is a nonlinear function of the reflectances (or transmittances) of the substrate and patterned layer materials, it is also important that the video system not subtract the background level from the signal used for measurement as is typically done to enhance the image for visual viewing or picture taking. If the reflectance of the materials changes (as is always the case when layer thicknesses change on wafers), automatic subtraction of the dc or background light level causes the edge detection threshold to shift or float, introducing a dimensional measurement error.

The response range in a video camera, which is limited by the saturated black and white levels, introduces a different problem. The light level in the image must be adjusted using lamp voltage or neutral density filters so that the edge detection threshold to be used for measurement lies on the linear portion of the response curve. If the lamp intensity level is incorrectly set, the lamp intensity level changes, or the reflectivity of the materials changes, again the edge detection threshold will change or float, and line-width errors will result.

In addition to the parameters discussed above that directly affect feature-size measurement, conventional imaging properties of video cameras must be considered in system design, such as image retention characteristics, stability of the electronics, and distortion.

In photometric systems (e.g., photomultiplier or photodiode) the desired linearity is generally achieved without problems, provided that the illumination level is adequate. However, the scanning slit width must be one-sixth or less of the Airy disk diameter at the image plane in order not to degrade the image, that is,

$$\text{Slit width} < \frac{M}{6}\left(\frac{1.22\lambda}{\text{NA}}\right) \tag{9}$$

where M is the magnification at the slit plane. The analog to slit width in a video camera is the camera line spread function. The effects of a Gaussian line spread function on image structure have been modeled by Kirk *et al.* [17].

It should be apparent from the above discussion of detectors that a video-based measurement system must be set up with greater care to avoid introducing serious metrology errors in feature-size measurements.

I. Vibration

The effects of vibration are clearly dependent on the time taken for a given measurement relative to the frequency, as well on the magnitude of the vibrations. The primary effects are to degrade the image profile and reduce the precision of the measurement. With desired precision of linewidth measurements at the level of 0.01 μm (100 Å) or less, some kind of vibration-isolation system is absolutely necessary. A site survey to determine the vibration characteristics that a given system will be subject to is highly recommended. This survey will help determine the properties required in the vibration-isolation system. In addition, the slower the measurement (e.g., a photometric measurement system is slower than a video-based system), the more massive the vibration-isolation system should be. While honeycomb construction eliminates many higher-frequency modes, the only effective damping for low frequencies is increased mass.

J. Effect of Materials on Image Structure

Materials encountered on IC photomasks and wafers fall into two groups, thin and thick layers, with the dividing line at approximately $\lambda/4$. This distinction arises out of the relative difficulty of theoretically predicting the waveforms produced by lines patterned in thin versus thick layers. The imaging of thin layers in microscopy can be treated using scalar theory of partially coherent imaging [5]. Early microscopy and metrology treated images of opaque line objects where the effects of thickness and phase were not considered. Although it was known that coherent and partially coherent imaging systems were sensitive to phase, the importance of the phase discontinuity at the line edge in determining the image structure and its effect on linewidth measurement is a relatively recent inclusion in the theory [13].

The image characteristics of thin layers of both dielectrics and metals were discussed previously [18]. At that time, it was recognized that thicker layers of oxide, resist, and metals produced image profiles that could not be predicted by scalar theory and the simple concepts of reflectance and phase retardation.

Since then, a model has been developed [13,19] that better predicts the microscope image waveforms for line objects patterned in thick layers. This theory essentially describes the coherent imaging of small three-dimensional line objects arranged in a periodic pattern. Assuming the line to be infinitely long and the dielectric constant to vary in only one dimension (i.e., constant with depth in a single layer), the dielectric constant ϵ is expanded in a Fourier series,

$$\epsilon(x) = |\eta_c|^2 = \sum_n e_n \cos(2\pi n x/P), \qquad (10)$$

where P is the period of the line–space structure, the e_n are the Fourier series coefficients, and η_c is the complex index of refraction. A solution is then found to the E-field wave equation (TE case),

$$\partial^2 E_y(x, z)/\partial x^2 + k_0\epsilon(x)E_y(x, z) = 0, \qquad (11)$$

in the form

$$E_y(x, z) = \sum_m \left[D_m \exp(\alpha_m z) + D'_m \exp(-\alpha_m z) \right] \sum_l B_{lm} \exp[ik_0(\lambda l/P)x]$$

$$(12)$$

where k_0 is the wave number in air, B_{lm} are the eigenvectors, and α_m the eigenvalue solutions of Eq. (11) representing "waveguide" modes. This method of solution for the E field has been used in waveguide analyses [20,21]. The difference here is in the treatment of the boundary conditions that determine the Fourier coefficients for the transmitted and reflected fields. To simplify the analysis, a single plane wave (coherent limit) is assumed to be normally incident to the surface of the patterned layer. Any multilayer composition of unpatterned sublayers may be included. Once the Fourier series components of the reflected (or transmitted) E field are found, the image formed in a microscope can be calculated using conventional methods [22]. The Fourier series expansion of the reflected field may be regarded as representing a pseudo-object that is "seen" by the microscope.

This model takes into account two major effects that cannot be accounted for by the traditional scalar theory of partially coherent imaging, namely, the multiple reflections that occur within a transparent layer (the same effect that occurs in ellipsometric measurements of layer thickness) and the waveguide effects that occur near the line edges for thick layers of both dielectrics and metals. In the latter case, the aspect ratios (layer thickness/linewidth) at micrometer and submicrometer dimensions are such that weak Bragg diffraction or mode coupling occurs, and this both distorts the image structure seen in the microscope and affects the algorithms needed for accurate edge detection.

Figure 15 compares image waveforms for thick and thin layers of dielectrics and metals. The image structure for thick layers is sensitive to both wavelength and angle of incidence; the thicker the layers, the more sensitive they are, especially to angle of incidence. In Fig. 15, the thick-layer calculations assume nearly normal incidence, corresponding to a highly coherent ($S < \frac{1}{3}$) bright-field imaging system. These calculations would

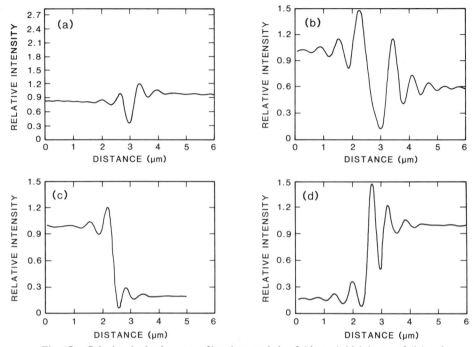

Fig. 15. Calculated edge image profiles characteristic of thin and thick layers of dielectric and metal showing difference in image structure. (a) Thin layer (140 nm) of patterned silicon dioxide on silicon; (b) thick layer (0.6 μm) of patterned silicon dioxide on silicon; (c) thin layer of metal; (d) thick layer (0.6 μm) of metal on silicon.

therefore not apply to a focused laser beam scanning system [7], where the scattered or diffracted field must be calculated as a function of angle of illumination and integrated over the cone angle of the illuminating objective. To date, these calculations have not been done owing to their complexity.

In addition to lines with vertical edges, this waveguide model has been extended to treat variable-geometry line cross sections, as shown in Fig. 16 [19]. The effects of different edge geometries on the image waveforms for a 0.6-μm-thick silicon dioxide layer are shown in Fig. 17. In most cases, nine layers appear sufficient to model these geometries. The model also demonstrates that, with sufficiently narrow spectral bandwidth and a narrow illuminating cone angle, submicrometer lines can be resolved (see Fig. 18).

It should be noted that this method of solution, while best suited for small objects and diffraction gratings, does not require any of the usual approximations found in the conventional treatment of rigorous diffraction theory (i.e., infinite conductivity, the Born approximation, or Ray-

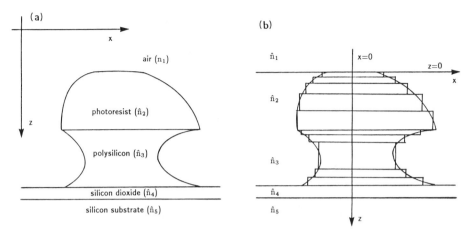

Fig. 16. Cross section of a typical line object patterned in thick layers (a) and the corresponding multilayer representation (b) used in calculating the image profile (b). (From Kirk and Nyyssonen [19].)

leigh approximation) and, therefore, is ideally suited to the wide variety of materials and geometries found in IC manufacture. The major limitation of this method of computing image profiles is the size of the matrices used in the computations. A 12-μm period at a wavelength of 0.53 μm requires ± 22 diffracted orders and a 90×90 complex matrix inversion [13].

In addition to changing our ideas about optical linewidth measurement, this theoretical development also affects conventional ideas of focus and focusing techniques. At micrometer and submicrometer dimensions, one can no longer talk about focusing at the top or bottom of the patterned layer. Rather, the illuminating field is scattered by the line object independent of the imaging system. The best that we can do is focus the imaging lens on the field as it exits from the top surface of the patterned layer. The objective lens then coherently images this field at whatever plane coincides with the object plane of the lens. If the object plane of the lens does not coincide with the top surfaces of the patterned layer, a phase term equivalent to defect of focus is added that limits the resolution of the imaging system. In a coherent system with small objects and phase present, the field may be varying so rapidly with respect to position along the optical axis that conventional formulas for depth of focus do not apply. In metrology, a new definition of focus tolerance is needed, one indicating the tolerable defocus for which no change in linewidth occurs (with the tolerance specified).

It appears, therefore, that insofar as resolution is concerned, the proper focus position is the top surface of the patterned layer. Although sensing

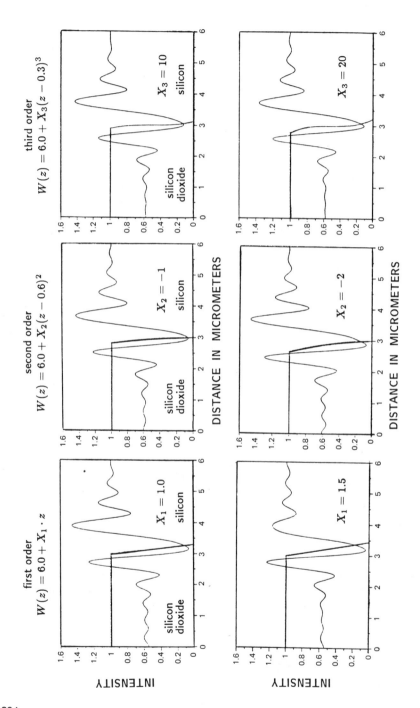

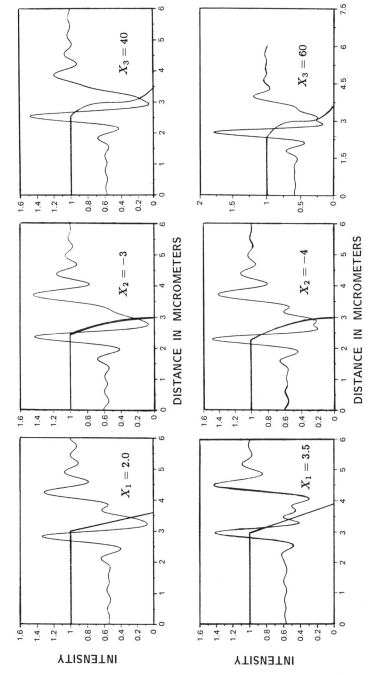

Fig. 17. Effect of edge geometry on the calculated image profile for patterned silicon dioxide (0.6 μm thick) on silicon. (From Kirk and Nyyssonen [19].)

295

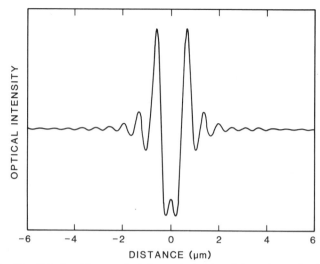

Fig. 18. Calculated image of a 0.5-μm window in a 0.6-μm layer of resist.

the top surface may seem to be an easier problem than determining best focus from the complex image structure for thick layers, in fact, all methods of optically sensing the top surface are affected by the multiple reflections occurring within the layers.

This discussion of optical imaging for metrology is by no means exhaustive. With the push to submicrometer feature size for VLSI, process control instrumentation is undergoing an evolution toward greater automation and sophistication. Undoubtedly, this discussion will require revision within a few years in light of new instrumentation and our improved understanding of the requirements for optical metrology.

IV. IMAGING IN THE SCANNING ELECTRON MICROSCOPE

Application of the SEM to feature-size metrology in IC applications is relatively recent in comparison with optical metrology. Although the SEM has been used for many years as a diagnostic tool, in many respects it is years behind optical metrology, particularly in terms of understanding the metrology aspects of the instrument. In order to take full advantage of the potential for increased resolution, procedures for instrument adjustment as well as accurate and precise calibration and measurement techniques must be developed.

Some of the comments of the previous section apply to SEM metrology as well as optical metrology. Just as the optical microscope as a metrology tool evolved from a visual inspection and picture-taking instrument, so too has the SEM. Image enhancement as defined earlier is frequently utilized in SEM inspection and picture taking and is just as disastrous to accurate feature-size measurement in an SEM as it is in an optical microscope.

Again, the requirements for detection and line-spacing measurements are much less demanding than for feature-size measurement. Because identical images of the lines are sufficient for accurate line-spacing measurements, a nonreproducible instrument adjustment can be tolerated for determining magnification. Again, feature-size measurement requires reproducible, known image profiles properly analyzed. With this in mind, we shall discuss some of the most important parameters of the SEM and their influence on metrology, leaving details of SEM technology and operation to the many available texts on the subject [23–25].

A. Resolution

The great attraction for SEM metrology has always been the extremely short wavelength associated with electrons and the potential for far superior resolution, of the order of 10 Å or less at 20 keV. In practice, such resolution is not realized because of inherent electron lens aberrations and random noises fluctuations due to the discrete nature of the electron beam [26]. Instruments today can achieve resolutions of 30 to 50 Å at optimum beam voltage and optimum working distance. At other working distances and low voltage, the beam diameter may increase to 200 Å or more. However, beam diameter is not the most serious limitation in SEM metrology. Effects due to the electron-beam interaction with the material currently appear to be the major limiting factor in the application of SEM metrology to IC manufacture.

B. Aberrations

Aberrations, including spherical and chromatic aberrations, are inherent in the design of electron optics and the dispersion (in energy) characteristics of the electrons. In addition, some astigmatism is usually present due to the imperfect machining of parts and imperfect alignment of the column. The SEM operator has traditionally been given the responsibility of correcting for the astigmatism based on visual evaluation of the image structure. Dirt, both particulate contamination and uniform coating of the system components, also enhances aberrations in the system [27].

In addition to axial spherical and chromatic aberrations, other off-axis aberrations are introduced by the scanning process, including coma and additional astigmatism components. At high magnification, where the field of view is limited, the latter contributions may be small compared with spherical aberration. Off-axis aberrations, of course, affect both line-spacing and feature-size measurements. The accuracy and uniformity of the raster scan may be the most important source of metrological errors off-axis. Unlike optical systems, in which the magnified image is scanned, thereby reducing the effects of scan nonuniformities, in the SEM the raster scan must be accurate to better than the overall precision desired of the measurement. State of the art SEM instrumentation is undergoing improvement here as the needs of accurate submicrometer metrology are being recognized.

The last of the aberrations to be discussed here, defocus, is of less concern because of the increased depth of field in an SEM. Typically, the depth of focus may vary between 1 μm and 1 mm depending on the magnification as well as the size of the final aperture and working distance. Focus has been traditionally adjusted by visual evaluation of the image on the cathode ray tube (CRT). With the development of more automated SEM linewidth measurement systems, many aspects of SEM operation are currently changing, including the development of autofocus techniques based on maximizing edge slope or, in the case of secondary electron imaging, maximizing the overshoot at the line edges.

In direct parallel to optical systems, sloppy column alignment, particularly the gun and final aperture, and system adjustment directly affect the reproducibility of waveforms and, therefore, of measurement accuracy and precision. Unlike optical imaging systems [11], however, there are at present no generally accepted procedures for adjusting and checking the alignment of the column and system operation to ensure reproducible waveforms. In fact, reproducible alignment and adjustment is a greater problem with an SEM because of the frequency with which the column is torn down for cleaning or source replacement.

As the SEM moves into production as a metrology tool, such procedures will have to be developed before the full potential of the instrument can be achieved. In addition, where adjustments significantly affect the image waveforms, as has happened with focus in the optical microscope, automated, objective, and operator-independent methods of adjustment will have to be developed.

C. The Detector

In an SEM, a voltage (current) output is derived from an electron detector and usually viewed on a CRT. The industry is gradually moving away from the crude metrology of yesterday, which used a ruler to take measurements off the CRT or a photograph. Now, as with some optical systems, the output signal from the SEM is being digitized with a computer used for data analysis, thereby eliminating all of the waveform and distance distortions introduced by CRT displays and photography. In addition, computer data analysis allows the use of more objective criteria for edge detection and measurement.

The problems with a CRT video display when its response is part of the measurement process were discussed earlier with respect to optical imaging using a video system and apply here as well. Similar problems occur when a photograph is used in the measurement process. The exposure time, nonlinear photographic response curve and the slope or γ associated with it, and limited density range corresponding to the black and white levels in a video system or CRT display all contribute similar distortions to the image. In addition, it is impossible to apply any objective criterion such as an edge detection threshold when using a ruler on a photograph.

One of the most important problems that finds its way into metrology is asymmetry of the line image [28]. Most electron detectors have a preferential axis either because of their orientation to the sample (other than normal) or because of their split design (e.g., quad detectors). Alignment of the line to be measured to this axis is critical to accurate metrology but is frequently difficult to determine visually due to the irregular nature of IC wafers. That is, asymmetry in the line structure may be partially compensated for by misalignment relative to the detector axis unless some independent method is used to define the detector axis.

D. System Backscatter

In SEM systems, there is a problem analogous to that of flare or stray light in an optical system, namely, system backscatter. In the two imaging modes most commonly used for IC inspection, secondary-electron emission and backscatter, there is always a sizable contribution of additional secondary electrons produced by backscatter electrons impinging on system components and specimen area outside the region of interest. The percentage of these additional electrons may vary and, in the case of secondary-electron imaging, has been estimated to be as high as 60%. This variation makes it difficult to model the imaging process and predict image structure or develop accurate and reproducible edge detection criteria. The amount is also dependent on such factors as operating voltage and tilt of the specimen.

In the process of contrast enhancement normally used for visual viewing (consisting of increasing gain and adjusting image brightness), a dc bias is effectively subtracted from the signal. However, as is the case with coherent optical imaging, the backscatter or secondary-electron emission coefficients (reflectance or transmittance in the optical case) of the patterned layer and substrate may vary from one wafer or structure to the next. The amount of system backscatter may vary, and the edge detection threshold may not be a simple linear function of these coefficients. In some cases, subtraction of a dc bias may allow the edge detection threshold in the image to wander with respect to the true edge location. When this occurs, a correction for change in materials or edge geometry cannot be made and metrology errors may go undetected.

E. Charging

The penetration of electrons into a specimen, particularly in the case of dielectrics such as resist and oxide, produces charging and thereby alters the field potentials, etc. Although this problem has been dealt with traditionally by gold or carbon coating of the specimen, this process is destructive to IC wafers and is therefore not acceptable for routine inspection and CD measurement.

In more recent years, there has been a trend toward low-voltage operation of SEMs [29], and most manufacturers have added or extended the capability of their SEMs to voltages as low as 0.5 keV or lower. In addition to eliminating the need for specimen coating, low voltage is attractive because it minimizes damage to IC devices [30].

However, the disadvantages of this mode of operation are the loss of resolution, increased susceptibility of the electron beam to local charging, and the difficulty of modeling the electron beam – material interaction and associated development of accurate edge detection criteria.

F. Beam Stability, Vibration, and Stray Fields

Another serious limitation on precision in dimensional metrology arises from instability in either the intensity or the position of the probe beam, variation in the collector field, or motion of the specimen relative to the probe beam such as produced by vibration. Because of the high magnification used, SEMs have been traditionally placed on some type of antivibration supports. Because of the numerous sources of vibration generally found in wafer fabrication areas and the desire for improved measurement precision, the problem is even more severe in IC applications. Of particular concern are low-frequency vibration from various sources including vacuum pumps in the SEM itself, acoustic sources of vibration, and stray electric and magnetic fields [31]. Although site surveys are generally performed at the time of installation in order to identify such problems, it must be kept in mind that in dimensional metrology applications the degree of isolation required is generally more severe. In addition, the findings of the site survey may become invalid due to changes in the environment (e.g., addition of new equipment). Isolation systems with better frequency damping characteristics and vibration isolation than are currently installed on most SEMs can be retrofitted for use in hostile environments. Additional shielding can also be added. The ultimate test, of course, is the reproducibility of both signal waveforms and measurements from one day to the next.

G. Electron-Beam Interaction with Materials

One of the most difficult tasks in discussing the requirements of SEM dimensional metrology is conveying the idea that what is seen in an SEM picture may not be an accurate representation of the feature being viewed. Figure 19a is a split-field picture of secondary and backscatter images of a chromium line on silicon [32]. Partway through the exposure the video display was shifted from the secondary to the backscatter signal and then back to the original secondary image. Since the line object itself has not changed its size or structure and the pitch has remained constant, the apparent difference in linewidth shown in Fig. 19b is due to the imaging process. By any criterion currently used for edge detection, the two images would not yield the same linewidth measurement. Which, if either, is right? What is the true linewidth?

The dilemma is analogous to that which occurred in optical measurements in the mid-1970s when reflected and transmitted light measurements yielded different linewidths on photomasks. We know today that both were in error. At that time, our understanding of the mechanism of coherent optical imaging was incomplete, and the edge detection criteria used were incorrect. Today, with well-understood waveforms and accurate edge detection algorithms, such measurements in bright-field imaging agree to within the uncertainty of either measurement technique, currently $\pm 0.05 \mu$m.

It is clear that we do not understand the mechanisms of electron optical image formation well enough quantitatively to develop accurate edge detection criteria. A reference standard that arbitrarily declares a line to be 1 μm wide based on a peak-to-peak signal measurement in a secondary-elec-

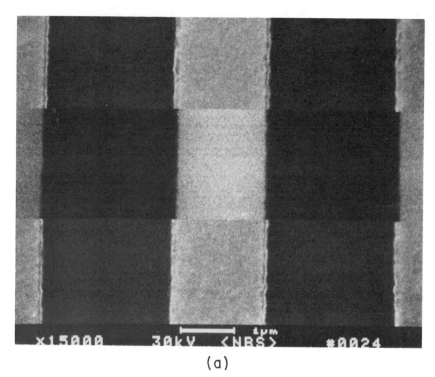

(a)

Fig. 19. (a) Split-field micrograph of secondary-electron (top and bottom) and backscattered-electron (middle) images of a chromium line on silicon; (b) waveforms showing pitch and linewidth measurements for corresponding secondary (left) and backscattered (right) images (15,000 ×, 30 kV). Measurement was done using an arbitrary 40% automatic threshold crossing algorithm.

tron image will not reconcile the difference between secondary and backscatter measurements or between optical and SEM measurements. In addition, in order to evaluate the precision and accuracy of a measurement, we have to have an accurate model to predict the image profile and determine the parameters of the line material and SEM system that influence the position of the peak or any other criterion relative to the true edge geometry. Only when models for secondary and backscatter imaging properly predict the relation of the image waveform to edge geometry and sample characteristics, thereby producing linewidth measurements that agree to the desired precision and accuracy expected of an SEM, will the SEM have achieved its full potential as a metrology tool.

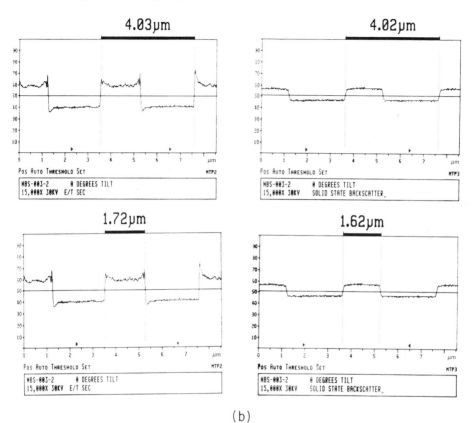

(b)

Fig. 19 *(Continued)*

H. Modeling

This chapter would not be complete without some discussion of the current level of modeling efforts. Two basic approaches have been taken: (1) Monte Carlo modeling and (2) a phenomenological approach.

Monte Carlo calculations [33] trace the paths of individual electrons incident on the specimen through elastic and inelastic scattering processes until all of the electron energy is lost or the electron escapes through the surface. Thousands of trajectories are needed at each point of impact of the electron beam as it scans to produce an image profile [34]. In order to make the problem tractable, the following assumptions are usually made. (1) Only elastic scattering events that appreciably change the electron's direction of travel are considered, and the distance between scattering events as the electron travels in the material is taken as the mean free path or multiple thereof; (2) at each event, the type of event (x-ray generation, backscatter, etc.) and scattering angle are chosen using random numbers; (3) the Bethe relation [35], which assumes continuous energy loss per unit distance, is used to calculate the energy losses between events; and (4) the line structures are typically modeled with trapezoidal cross sections with a given edge slope angle, as shown in Fig. 20.

Monte Carlo modeling was originally developed for high-voltage systems and x-ray microanalysis, not for linewidth measurement. Nevertheless, it clearly gives a good approximation of the electron-beam interaction behavior and the image profile produced. However, some of the limitations of the current level of modeling include (1) overly simplistic assumptions that break down for low voltage and very thin layers, (2) inability to treat the interface region in multilayer structures, (3) inability to treat more complex geometries such as those shown in Fig. 16, (4) inability to treat small geometries where the edges are sufficiently close together that the adjacent

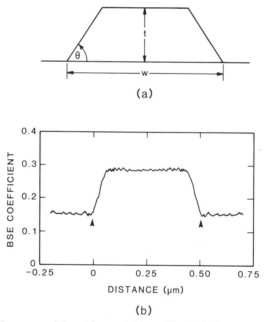

(a)

(b)

Fig. 20. (a) Geometry of line object and (b) resulting SEM image calculated by Monte Carlo method (from Jensen *et al.* [34]). Monte Carlo simulation of the backscattered-electron (BSE) image produced by the scan of a 20-keV electron beam across a chromium line on a silicon substrate with an edge slope of 70°. The arrows denote the material line edge positions.

edge influences the image structure (submicrometer linewidths), and (5) inability to include the effects of charging of the specimen. Some of these limitations can easily be overcome by relatively simple extensions of the existing models, for example, extension to more complex geometries and smaller linewidths. Other limitations are more fundamental and will require major advances, for example, treatment of low-voltage beam interactions and charging.

The *phenomenological approach* to modeling attempts to find an equation(s) with variable parameters that fit the experimental data. Such a model was developed by Swing [36] for secondary-electron imaging assuming a Gaussian beam incident on the geometry shown in Fig. 20. Different secondary-electron emission coefficients were assumed for the materials on either side of the edge, and the response at the inclined portion of the edge was assumed to increase by sec θ [37]. In this model, the Gaussian beam was assumed to have a diameter of the order of, or larger than, the edge width, and therefore its characteristics dominate the image waveform. The model also assumed no electrons scattered beyond the region of beam impact.

Phenomenological models are useful in that edge detection algorithms can be easily developed from them. However, the parameters of the model must be selected either from known physical data or by empirical fitting of the measured image profile. This approach to modeling should be explored further in order to assess its usefulness and adaptibility for routine metrological work.

V. DIMENSIONAL METROLOGY

In dimensional measurements of microfeatures, it is not possible to use contact methods to make measurements. Instead, the measurement plane is removed in space from the actual feature(s) being measured. Therefore, early in this chapter we introduced the idea of two coordinate systems, one in the desired measurement plane, the other in the actual plane of the measurement (e.g., an image plane of the microscope). The measurement system must accurately and reproducibly map one onto the other. The discussion of the imaging portion of a dimensional measurement system (optical and SEM) that followed included many possible sources of error in the mapping from object to image. In the following sections, we look at possible sources of error in the subsequent measurement process, either in scanning the image plane or in measurement of distance in the object plane by moving the stage.

Establishing Coordinate Systems

The coordinate system in the plane of the measurement is usually the easiest to define. It can be defined at or near the feature of interest on the wafer or photomask. For long-distance measurements between features, the choice becomes somewhat more complicated; leveling and chucking of the wafer or photomask have to be considered. The primary purpose of leveling (tilt adjustment) and chucking is to ensure that the surface of the wafer is coincident with the focal plane of the measurement system; that is, both edges or features must be in focus. The longer the dimensions to be measured and the higher the NA of the imaging lens, the more accurate the leveling and chucking must be. The problem is analogous to that of focusing in a wafer stepper, except that we have the option here of focusing and leveling for each measurement as well as for each die and globally for the entire wafer. Compared with a wafer stepper, however, the tolerances are tighter because there is less depth of focus due to coherence, wavelength, and the high NAs used and because of the requirement that the measurement gauge be 3–10 times more accurate than the instrument that produced the part.

In summary, over short distances it may be sufficient to level the mask or wafer holder, whereas over long distances, chucking as well as leveling of individual masks or wafers at each measurement site may be required. Next, we shall consider the definition of the coordinate system in the actual measurement plane.

Example 1. Image Plane Measurements over Long Distances. Let us assume for the moment that the wafer is chucked and leveled and a coordinate system (x, y, z) can be defined on the surface of the wafer or photomask, at or near a pair of lines whose center-to-center spacing is to be measured. Let us assume that the lines are imaged onto the plane of a crosshair or filar in a microscope, which here defines the image plane. Defining a coordinate system (x', y', z') in the image plane at the crosshair or other fiducial mark seems fairly straightforward. Next, we must map or relate these two coordinate systems (x, y, z) and (x', y', z').

The usual assumption is that, in the direction of measurement,

$$x' = Mx + x_0, \tag{13}$$

where M is the magnification and x_0 the zero position on the distance measurement scale. However, such an ideal system is unrealizable. Numerous sources of error may be present in addition to those already discussed. The image plane may not be flat and may coincide with the measurement axis only at one (or more) point(s); there may be vibration so that x_0 becomes a random function of time; and the detector and mechanism used to measure distance, (e.g., filar eyepiece), scanning slit with lead screw and shaft encoder, or video system may introduce other errors.

In general, we can write

$$x' = Mx + x_0 + \xi(x, y, z, t, \theta), \qquad (14)$$

where ξ is an error term that may be a function of position coordinates, time, and the alignment error angle θ between the line(s) to be measured and the measurement axis. In addition, some components of the error are not reproducible (systematic) and must be treated by statistical averaging. For example, the alignment of the feature to be measured to the measurement axis (θ) is never reproduced exactly, nor is the alignment of the system and the position of the feature in the field of view, so that contributions to the error from aberrations vary.

In accurate metrology, the objective is to reduce the total uncertainty U as defined in Eq. (3) below the level of the desired measurement accuracy, that is, one-third to one-tenth of the tolerance on the part being measured. In the present case, we are concerned with measurement errors as small as 0.01 μm, which are easily found in real systems due to manufacturing tolerances on such things as optical components, column alignment, and lead screw errors. Therefore, U may be significantly larger than desired or indicated by the manufacturer's quoted precision and must be determined through calibration with a known reference standard. As discussed earlier, both the standard (materials and linewidth or line-spacing dimensions) and the sampling method used for taking measurements (number of measurements and time period) must match the intended use of the measurement system. Contributions to the total uncertainty U from both the random error s and any systematic error E are determined from repeated measurements on a reference standard [2], removing and inserting the photomask or wafer for each set of measurements. When the value of U over the dimensional range of interest exceeds the desired measurement tolerance, it must be determined whether the measurements are reproducible enough to apply a correction to reduce the systematic error and thereby the total uncertainty. In general, correction of measurements using a calibration curve is recommended only when test statistics such as described in Croarkin and Varner [2] indicate the existence of a truly systematic and reproducible error. The total uncertainty U can be substantially reduced only when the precision as indicated by three times the standard deviation is substantially less than the systematic error E.

In addition, a reference standard containing a small number of approximately equally spaced line pairs is useful only for correcting linear calibration curves. In general, a nonlinear calibration curve as in Fig. 9b requires more data points unequally spaced with a concentration of data points in

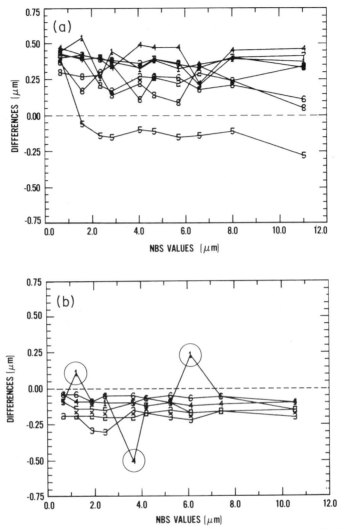

Fig. 21. Some examples of common problems found in calibration data (for linewidth measurements, taken from Jerke *et al.* [39]). (a) One-day loss of system control, (b) outliers (circled), (c) precision too poor to correct for offset, and polarity-dependent errors in linewidth for (d) lines and (e) spaces. Line and space widths may not add up to correct pitch if polarity-dependent errors are not of equal magnitude (opposite sign) and system is improperly adjusted or corrected.

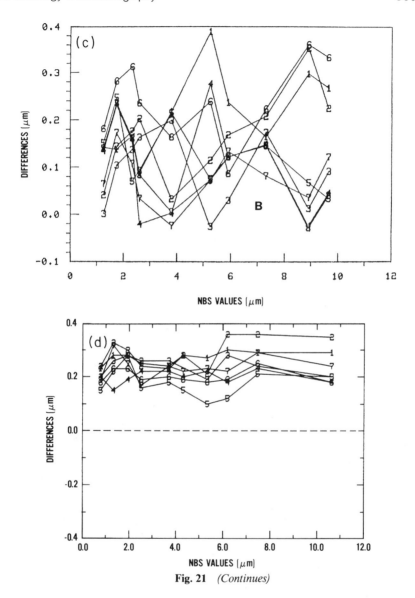

Fig. 21 *(Continues)*

the region of greatest curvature. Some examples of calibration data exhibiting different types of behavior are shown in Fig. 21.

Example 2. Object Plane Measurements of Distance. In general, if the distance to be measured exceeds the usable field of view of a reasonably high NA objective, a measurement system is used that translates the wafer

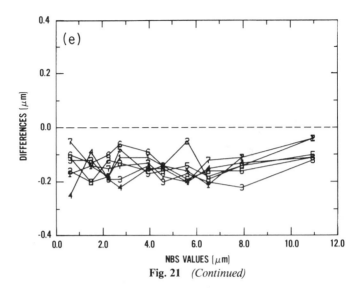

Fig. 21 *(Continued)*

or photomask and thereby transfers the actual measurement coordinate system to a measurement gauge in the object plane of the microscope. Usually, the distance moved is determined by a shaft encoder, stepping motor, or interferometer. The advantages of this approach are that all of the aberrations of the microscope imaging system and errors introduced by the detector are eliminated since the imaging system is used only axially (if properly aligned) to provide a fiducial mark on the wafer. On the other hand, the requirements on the precision motion of the stage as well as the ability to hold focus become much more severe.

For this example, let us assume that we have a wafer sitting on a moving stage with a plane mirror interferometer measuring the motion of the stage. Because the mirror must be displaced from the optical axis to accommodate the objective and sample, and the interferometer must be displaced far enough to accommodate the longest desired measurement distance (up to 10 in. on wafers), numerous sources of error are introduced, including errors due to roll, pitch, and yaw of the stage relative to the interferometer axis; a cosine error due to misalignment of the line object to the interferometer axis; Abbey offset error if the interferometer axis does not intersect the optical axis of the microscope; as well as vibration (from acoustic as well as mechanical sources) and other environmental effects.

For short distances (~ 1 mm or less) and short interferometer dead path (the distance between the interferometer and the moving reflector), temperature and humidity affect the stability of the measuring instrument rather than the part being measured, so that the interferometer least count and vibration may limit the accuracy. Thus, the quality of the interferometer parts and the stability of the mechanical design are of prime importance.

For longer distances, dominant sources of error are temperature effects (expansion and contraction of parts) on both the instrument and the part being measured. In addition, temperature and humidity fluctuations over the dead path of the interferometer (when measurements are made in air) become significant, so that environmental control becomes extremely important [38].

Example 3. Feature-Size Measurement: Edge Detection. In addition to all of the above sources of error that affect any distance measurement, feature-size measurement introduces additional problems, as already noted. The primary source of error is due to inaccurate edge detection. Most optical microscope- and SEM-based systems use a fixed threshold or other criteria or allow user-selectable edge detection criteria.

For optical linewidth measurements, because of the material dependence, as discussed earlier, there is no single edge detection criterion applicable to most systems that is accurate for all waveforms. Nearly opaque photomasks viewed in transmitted light can be measured using an edge detection threshold that yields the proper linewidth dimension and is selected by adjustment to a reference standard that produces the same edge waveform. Such measurements are least affected by small changes in the thickness of the mask material or surface contamination. However, edge geometry and roughness do produce observable changes in the edge waveform and result in measurement errors.

See-through photomasks and wafers, however, have waveforms that are extremely sensitive to small changes in thickness and index of refraction of the patterned layer and sublayer(s) in addition to edge geometry, edge roughness, and surface contamination. When a fixed edge detection criterion is used, such as those illustrated in Fig. 22 for thin layers on wafers (<200 nm), the measurement error varies with the parameters R and $\cos \phi$, as shown. A consequence of the lack of compensation for changes in material parameters (which results from the use of a fixed criterion) is that variations in thickness, index, and edge geometry that occur in normal wafer fabrication processes, even on a single patterned layer, will produce measurement errors that go unobserved despite normal calibration and control procedures based on a single reference standard, even when the material matches that of the part being measured. These errors referred to as variable offsets have been documented [40] and may have magnitudes of 0.3 μm or more.

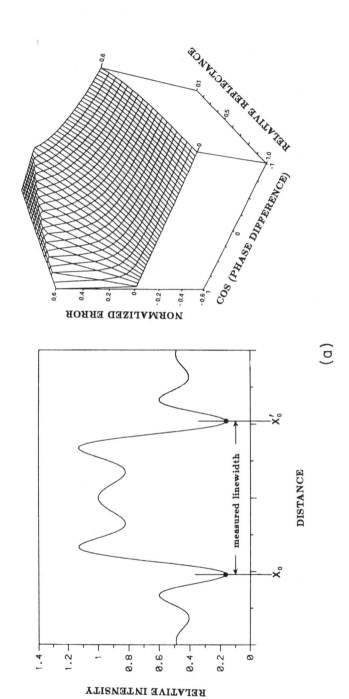

(a)

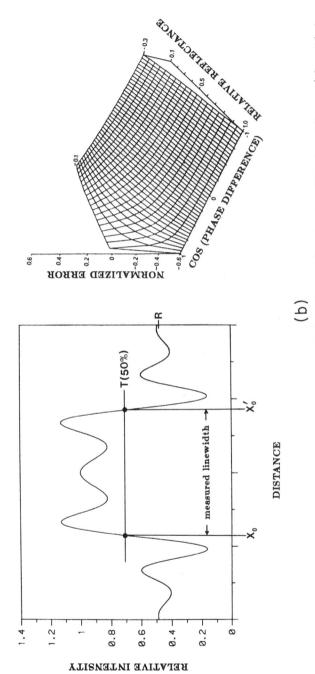

(b)

Fig. 22. Examples of normalized linewidth error as a function of relative reflectance R and cosine of the phase difference ϕ from thin-layer model for the edge detection criteria shown. (a) Minimum; (b) 50% threshold; (c) dual threshold. Linewidth error in micrometers for a coherent, diffraction-limited optical system can be found by multiplying the normalized error by λ/NA. (From Nyyssonen [14].) *Figure continues.*

313

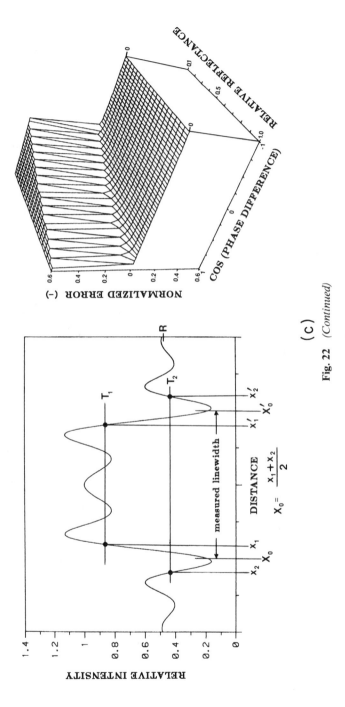

Fig. 22 *(Continued)*

Although opaque photomasks can be measured in either transmitted or reflected light, it is because of this increased sensitivity of reflected light measurements to material characteristics and geometry that the use of transmitted light is recommended [6]. The NBS photomask SRMs are designed for use in transmitted light and will not necessarily match the waveforms of other photomask materials when viewed in reflected light. Linewidth calibration errors will result from incorrect use of this SRM.

In SEM measurements, similar variations occur that are dependent on the characteristics of the materials (atomic number, secondary-electron or backscatter coefficients), edge geometry, surface contamination, beam penetration depth, etc. At this time, modeling of the electron-beam interaction with the materials of interest has not progressed far enough either to develop accurate edge detection criteria or to assess the magnitude of such errors. Noise or other sources that degrade or distort the image waveform also introduce errors. Both a poor signal-to-noise ratio (from inadequate signal level and poor detector sensitivity) and vibration reduce edge slope, perhaps changing the effective edge detection threshold and resulting in a loss of edge detection sensitivity.

VI. CONCLUSIONS

This chapter has attempted to provide a framework for realistically evaluating the capabilities of a given IC dimensional measurement system in the context of the purpose for which the measurement system is to be used. With current metrology tools, there appears to be no one best system suitable for all types of materials and levels of accuracy required. Many systems do specific metrology tasks well when the procedures used ensure reproducible waveforms and performance and when adequate control chart methods are used to monitor the accuracy and precision of the measurements and to ensure that the system remains in control within the desired limits.

As indicated, the requirements for distance or line-spacing measurement (for overlay measurement) and those for linewidth measurement are different: Reproducible magnification and line-spacing measurements do not guarantee either accurate or reproducible linewidth measurements. In addition, both optical and SEM measurements are material dependent so that both the precision and offsets or systematic errors in linewidth measurement are dependent on the characteristics of the materials, including thickness and edge geometry. Therefore, precision and offsets determined from a single reference standard may not guarantee process control.

When statistical evaluation indicates that the system is out of control or that the system has an unacceptably large systematic error or standard deviation, the difficult task remains of using the available data or devising additional tests to determine the largest source(s) of error, correcting, and retesting until satisfactory performance is achieved. Much of the discussion in this chapter can be used for diagnosing problems. There is such a variety of imaging systems, both optical and SEM, and mechanical designs, and variations are being introduced so rapidly, that no recipe can be given for this process of establishing realistic statements of precision and accuracy or for the process of bringing an instrument into control at the desired tolerance level when performance is unsatisfactory.

ACKNOWLEDGMENTS

The material in this chapter has been developed from the teaching of a large number of seminars and from interactions with people too numerous to mention. However, the author wishes to thank Bob Larrabee and Mike Postek for their helpful discussions, which have helped to clarify many of the points discussed here. Marilyn Dodge is also acknowledge for her assistance in acquiring and analyzing some of the data.

REFERENCES

1. "ASTM Manual on Presentation of Data and Control Chart Analysis," ASTM Spec. Publ. 15D. Am. Soc. Test. Mater., Philadelphia, Pennsylvania, 1976
2. C. Croarkin and R. N. Varner, Measurement assurance for dimensional measurements on integrated-circuit photomasks, *NBS Tech. Note (U.S.)* No. 1164 (1982).
3. NBS SRM 474/5, avail. from Off. Stand. Ref. Mater., Rm. B311, Chem. Bdg., Gaithersburg, MD 20899.
4. L. C. Martin, "The Theory of the Microscope." Blackie, London, 1966.
5. M. Born and E. Wolf, "Principles of Optics," 6th Ed., pp. 518–532. Pergamon, Oxford, 1980.
6. T. Wilson (with contributions from C. Sheppard), "Theory and Practice of Scanning Optical Microscopy." Academic Press, London, 1984.
7. D. Nyyssonen, *Proc. SPIE* **565**, 102–107 (1985).
8. D. N. Grimes, (aka Nyyssonen), *J. Opt. Soc. Am.* **61**, 870–876 (1971).
9. D. Nyyssonen, *Proc. SPIE* **194**, 34–44, (1979).
10. L. W. Smith and H. Osterberg, *J. Opt. Soc. Am.* **54**, 595–598 (1964).
11. "Standard Practice for Preparing an Optical Microscope for Dimensional Measurements," F728-81, Annual Book of ASTM Standards, 1981.
12. W. J. Smith, "Modern Optical Engineering," pp. 291–292. McGraw-Hill, New York, 1966.
13. D. Nyyssonen, *J. Opt. Soc. Am.* **72**, 1425–1436 (1982).
14. D. Nyyssonen, *Opt. Eng.* **26**, 81–85 (1987).

15. D. Nyyssonen, *Proc. SPIE* **480,** 65–70 (1984).
16. D. Nyyssonen, Linewidth calibration for bright-chromium photomasks, *NBS Intern. Rep. (U.S.)* No. 86-3357 (1986).
17. C. P. Kirk, D. S. Moore, and J. C. C. Nelson, *Opt. Eng.* **24,** 650–654 (1985).
18. W. M. Bullis and D. Nyyssonen, *in* "VLSI Electronics: Microstructure Science" (N. G. Einspruch, ed.), Vol. 3, pp. 301–346. Academic Press, New York, 1982.
19. C. P. Kirk and D. Nyyssonen, *Opt. Microlithogr. IV, SPIE Semin. Proc.* **538,** 179 (1985).
20. C. B. Burckhardt, *J. Opt. Soc. Am.* **56,** 1502–1509 (1966).
21. F. G. Kaspar, *J. Opt. Soc. Am.* **63,** 37–45 (1973).
22. E. C. Kintner, *Appl. Opt.* **17,** 2747–2753 (1978).
23. O. C. Wells, "Scanning Electron Microscopy." McGraw-Hill, New York, 1974.
24. J. I. Goldstein, D. E. Newberry, P. Echlin, D. C. Joy, C. Fiori, and E. Lifshin, "Scanning Electron Microscopy and X-Ray Microanalysis." Plenum, New York, 1981.
25. C. W. Oatley, "The Scanning Electron Microscope. Part 1: The Instrument." Cambridge Univ. Press, London and New York, 1972.
26. See Ref. 25, p. 22.
27. P. Echlin, *Proc. IITRI, 1975:* "Scanning Electron Microscopy," p. 679 (1975)
28. P. E. Russell, *in* "Electron Optical Systems," pp. 197–200. SEM, Inc. Chicago, Illinois, 1984.
29. M. T. Postek, *in* "Scanning Electron Microscopy" (J. J. Hren, F. A. Lenz, E. Munro, and P. B. Sewell, eds.), pp. 1065–1074. SEM, Inc., Chicago, Illinois, 1984.
30. W. J. Keery, K. O. Leedy, and K. F. Galloway, *Proc. IITRI, 1976:* "Scanning Electron Microscopy," pp. 507–514 (1976).
31. J. B. Pawley, *Scanning* **7,** 43–46 (1985).
32. D. Nyyssonen and M. Postek, *Proc. SPIE* **565,** 180–186 (1985).
33. K. F. J. Heinrich, D. E. Newbury, and H. Yakowitz, eds., Use of Monte Carlo calculations in electron beam microanalysis and scanning electron microscopy, *NBS Spec. Publ. (U.S.)* No. 460 (1976).
34. S. Jensen, G. Hembree, J. Marchiando, and D. Swyt, *Proc. SPIE* **275,** 100–108 (1981).
35. See Ref. 24, p. 72.
36. R. E. Swing *et al., in* "Semiconductor Measurement Technology" (W. M. Bullis, ed.), *NBS Spec. Publ. (U.S.)* No. 400–45, pp. 14–17 (1980).
37. See Ref. 25, p. 162.
38. J. S. Beers, A gage block measurement process using single wavelength interferometry, *NBS Monogr. (U.S.)* No. 152 (1975).
39. J. M. Jerke, M. C. Croarkin, and R. J. Varner, Semiconductor measurement technology, interlaboratory study on linewidth measurements for antireflective chromium photomasks, *NBS Spec. Publ. (U.S.)* No. 400–74 (1982).
40. P. Grant and T. Fahey, *Proc. SPIE* **480,** 30–32 (1984).

Chapter 8

Electrical Measurements for Characterizing Lithography

CHRISTOPHER P. AUSSCHNITT

GCA Corporation
IC Systems Division
Bedford, Massachusetts 01730

O happy he, who still renews
The hope, from Error's deeps to rise forever!
That which one does not know, one needs to use;
And what one knows, one uses never. — Goethe [1]

319

I. INTRODUCTION

Lithography, in the context of integrated circuit (IC) manufacture, is the critical step in the creation of the wafer patterns that are the floor plans, so to speak, of the multistoried circuit architecture. As pattern feature sizes and overlay tolerances shrink to keep pace with the increase in circuit complexity, there is a concomitant need to advance the methods used to characterize the performance of lithographic instruments. The limitations of optical and electron microscopy as applied to lithographic evaluation have driven more than one IC production engineer to summon the devil. Since about 1980, however, the electrical probe technique has emerged as a more effective and less costly ally.

Electrical probing can be applied to many different stages of the IC fabrication process, up to and including the finished IC itself. For our purposes, however, we are concerned with two major issues: (1) the fidelity with which a designed feature size on any given pattern level can be replicated on the wafer surface and (2) the accuracy with which any two pattern levels can be positioned with respect to one another. These imagery and overlay characteristics determine the ultimate density and complexity of the IC that can be fabricated using a given lithographic approach. Thus, they are the source of meaningful figures of merit for lithographic performance.

The imagery and overlay performance of lithographic systems determine the electrically measurable parameters, linewidth and registration, at each point on the wafer surface. The demands placed on measurement precision are driven by IC design rules. For state of the art circuits, a feature size of the order of 1.0 μm must be controlled to within 5% over the wafer, despite the fact that it is a function of numerous lithographic and process parameters. Likewise, an overlay budget of 0.25 μm is the net result of the minimization of many components — alignment, magnification, distortion, etc. — each of which may amount to only a small percentage of the total. In order to resolve the various factors contributing to imagery and overlay, measurement precision must approach 0.01 μm for both linewidth and registration.

In addition to precision, full characterization of lithographic performance requires that many measurements of linewidth and registration be made at each die location over the area of the silicon wafer on many successive wafers. The number of measurements rapidly mounts into the thousands. The acquisition and analysis of such quantities of data dictate the use of automated test procedures, mass data storage, and computers. The correct interpretation of results requires the application of mathematical models that accurately represent the physical mechanisms underlying the imagery and overlay performance of a given lithographic system. Furthermore, the use of statistical methods and a variety of display formats, such as histograms and maps, serves to pinpoint sources of error.

The electrical probe technique provides the necessary measurement precision and automation. The basis of the technique for lithographic evaluation is the design, fabrication, and probing of circuit patterns with electrical properties sensitive to linewidth and registration. Each measurement can be made with a total precision of better than 0.5%. The movement of the probes and the electrical data acquisition are governed by a computer controller in such a way that the system can sample hundreds of circuits in a matter of minutes. Data storage on magnetic disk makes possible subsequent analysis and display using standard software packages.

II. LITHOGRAPHIC SYSTEMS

A. Basics

In order to characterize lithography we must have a good grasp of the fundamentals of lithographic systems. A bewildering cast of competing systems has emerged since the early days of hand-cut rubylith and contact printers. The current stars of IC production are optical projection printers. Electron-beam, ion-beam, and x-ray systems, long billed as coming attractions, still wait in the wings. It is increasingly clear that photolithography will dominate for the near future. Advances in optical technology, coupled with multilevel process capability, will shepherd manufacturing into the submicrometer domain. Thus, we concentrate our discussion here on the properties of optical projection systems.

Optical projection lithography is the means of imaging the pattern of a master plate, a mask or reticle, in a photosensitive film on the surface of a wafer, as illustrated in Fig. 1. The projection system must also align the pattern precisely with respect to an existing pattern on the wafer. Hence, such systems are often called projection aligners. Projection aligners currently in use for IC production fall into two major categories: scanners [2–4] and steppers [5–7]. Their respective operation as seen by the wafer is shown in Fig. 2. The scanner paints the pattern over the wafer in a single stroke. The stepper applies the pattern in a serpentine series of exposures.

Scanners move the mask and wafer simultaneously. At any instant during the scan only a narrow illuminated region of the mask is imaged on the wafer. Over a full scan the illumination and image sweep across the entire mask and wafer, respectively. Alignment is restricted to a full-wafer,

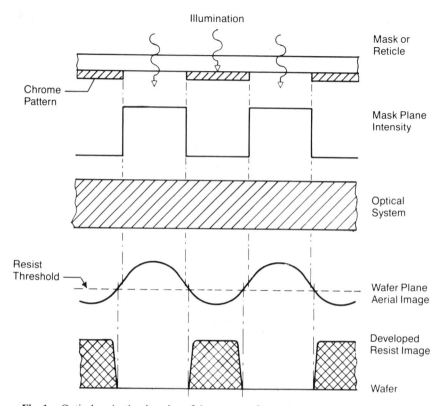

Fig. 1. Optical projection imaging of the pattern of a mask or reticle in a photoresist film on the surface of a wafer. In the case shown the resist is overexposed with respect to the mask pattern.

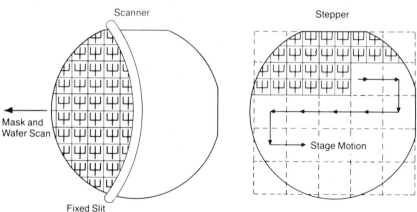

Fig. 2. Differences between scanner and stepper operation.

or "global," alignment of the mask to the patterned wafer. The mask used in the scanner, which is inherently a $1:1$ system, must contain all of the pattern information to be transferred to the wafer. As wafer sizes increase and feature sizes decrease, this places an increasing burden on mask-making capability. The reflective optics of the scanner make it a broadband system whose exposure spectrum is readily extended to deep-UV wavelengths.

The stepper takes advantage of the fact that the typical patterns consist of a repetitive array of fields corresponding to individual circuits or clusters of circuits. The reticle need contain the pattern of only a single field. The image of the reticle is then stepped repeatedly over the surface of the wafer to compose the full pattern array. Both global and field-by-field alignment to an existing pattern are allowed. The restricted field size facilitates the use of reduction optics in the stepper. Currently available systems have $1:1$, $5:1$, and $10:1$ reduction ratios. While the use of refractive optics in the stepper makes the move to shorter wavelengths difficult, the refractive optics and the compact exposure field of the stepper permit the achievement of high numerical apertures.

Both the scanner and the stepper are complex electro-opto-mechanical systems. We now consider in greater detail how the performance of each system is evidenced in the patterns it prints.

B. Imagery

By "imagery" we mean all of the qualities of the lithographic system necessary to produce patterns in photoresist with the desired minimum feature size, dimensional control, and profile over the wafer. The characterization of lithographic imagery has long been a source of confusion. Optical designers like to talk about "resolution" and "depth of focus," whereas the users of lithographic systems, IC process engineers, like to talk about "critical dimension control" and "process latitude." We shall show here that precise linewidth measurement capability, such as that provided by electrical probing, is the experimental bridge between the designer and user camps.

Discussions of photolithographic imagery invariably begin with the simple relations for the resolution or minimum feature size W_{min} and the depth of focus Z of a diffraction-limited optical system [8]:

$$W_{min} \propto \lambda/NA,$$
$$Z \propto \lambda/(NA)^2. \tag{1}$$

where λ is the wavelength of the exposure light and NA the numerical aperture of the optical system. The constants of proportionality are dependent on the particular procedure used to determine resolution and depth of focus. The above expressions provide a useful guideline in assessing the relative performance to be expected from different photolithographic systems. What is not disclosed by Eq. (1) is that, for a given lithographic system, feature size and depth of focus are coupled.

The key to a more thorough understanding of imagery is to relate the fundamental properties of the system to a measurable quantity, namely, linewidth [9,10]. For the sake of simplicity, let us restrict our discussion to the case of a periodic equal line and space pattern on the mask or reticle imaged in positive photoresist, as shown in Fig. 1. The imagery performance of a lithographic system operating near its resolution limit is dominated by the fundamental spatial frequency of the periodic pattern. The photoresist on the wafer acts as a threshold detector; that is, for a fixed development process it clears from the wafer surface at a fixed exposure.

Under these conditions, we can derive a simple expression for the dependence of the printed linewidth W on the exposure dose E and the defocus z of the image plane (see Appendix A). For the incoherent optical system characterized by a modulation transfer function $M(W_0, z)$, we obtain

$$\frac{W}{W_0} = \frac{2}{\pi} \cos^{-1} \left[\frac{\pi}{4} \frac{1 - E_0/E}{M(W_0, z)} \right], \tag{2}$$

where W_0 is the conjugate linewidth and E_0 the exposure dose required to print $W = W_0$. Note also that E_0 corresponds to the exposure energy threshold of the resist.

The well-known dependence of the incoherent modulation on W_0 is shown in Fig. 3. The point at which the modulation goes to zero, $W_0 = 0.25\lambda/\text{NA}$, determines the ultimate resolution of the optical system. In practice, the resist characteristics limit the usable modulation to the regime $1.0 > M \gtrsim 0.5$. As indicated in Fig. 3, the modulation is also dependent on the distance z of the wafer surface from the image plane of the optical system. In the regime of interest, the modulation transfer function can be approximated by [10,11]

$$M(W_0, z) \cong 1 - \frac{0.3\lambda}{\text{NA}} \frac{1}{W_0} - \frac{(\text{NA})^3}{3\lambda} \frac{z^2}{W_0}. \tag{3}$$

At $z = 0$, the smallest printable feature is given by setting $M = 0.5$, in accordance with the resist limitations, to obtain $W_{\text{min}} = 0.6\lambda/\text{NA}$. Under the above assumptions, however, a feature of size W_{min} would have zero depth of focus.

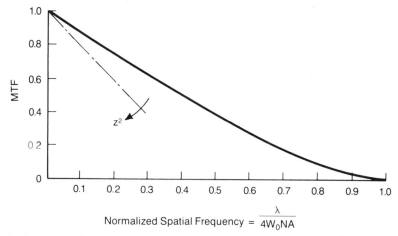

Fig. 3. Modulation transfer function (MTF) for an incoherent optical system. In the high-modulation regime the incoherent MTF decreases with the square of the defocus of the wafer plane.

Substitution of Eq. (3) in Eq. (2) enables us to calculate the exposure–focus characteristics of incoherent lithographic systems. The calculated curves of Fig. 4 give an immediate picture of the exposure and focus latitude available for a desired linewidth control. Although the focus latitude of the scanner is larger than that of the stepper, the imagery of the stepper is superior to that of the scanner. In fact, the dashed lines of Fig. 4 point out that the imagery of the 1.5-μm line at optimum focus on the scanner is only equivalent to the 3.5-μm out-of-focus imagery of the stepper! This is a dramatic portrayal of the virtue of a higher numerical aperture system.

The requirements on exposure and focus latitude for IC manufacture are determined by the characteristics of the wafer surface (topography and reflectivity variations) as well as the lithographic system control. Thus, the exposure–focus curves also give a measure of the allowed process latitude.

Many simplifying assumptions were used to obtain the expressions in Eqs. (2) and (3). In particular, we have ignored the effects of aperiodic patterns, spatial frequencies higher than the fundamental, partial coherence, lens aberration other than defocus, and process effects such as lateral resist development. More complete computer analyses have been described elsewhere [12,13]. The problem with computer models is that they tend to obscure the underlying physical principles. The above equations provide the necessary guidance for a strategy to characterize imagery performance using linewidth measurement. The ability to measure linewidth with speed

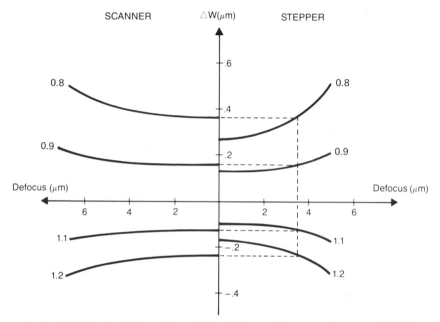

Fig. 4. Calculated variation of linewidth with focus and exposure for a nominal 1.5-μm line on both a scanner and a stepper. The dashed lines indicate that the imagery of the scanner at optimum focus is only as good as that of the stepper 3.8 μm out of focus. Scanner: $\lambda = 0.4$ μm; NA $= 0.17$. Stepper: $\lambda = 0.436$ μm; NA $= 0.3$.

and precision provided by the electrical probing technique enables us to generate experimentally the characteristic curves shown in Fig. 4. It is the experimental curves that specify the imagery performance of a lithographic system for a given feature size at a given point on the wafer surface.

Full characterization of imagery requires linewidth measurement at an array of sites distributed over the printed area. In the case of the scanner we need concern ourselves only with a representative sampling over the wafer. For the stepper, we must distinguish between intrafield and interfield measurements. Intrafield measurements determine the uniformity of lens and illuminator performance over the field. Interfield measurements determine the repeatability of exposure and focus from field to field.

Both scanner and stepper require that we characterize imagery in at least two orthogonal orientations. For the scanner we expect imagery differences between orientations that are parallel or perpendicular to the curved slit. Within the field of the stepper we expect differences between radial (sagittal) and circumferential (tangential) orientations with respect to the optical axis (usually the center of the field). The measurement of these differences over the wafer, in the case of the scanner, and over the field, in the case of the stepper, provides an important diagnosis of the optical quality of each system.

C. Overlay

As in the case of imagery, overlay characterization of a lithographic system requires that we relate fundamental properties of the system to a measurable quantity. The measurable overlay parameter is the pattern registration at any point on the wafer with respect to an adopted standard. For measurement systems based on an interferometric laser stage, the pattern position is relative to the position determined by the interferometer. For electrical measurements, the pattern position is determined relative to a previously printed pattern. It must be kept in mind that the measurements of position are accurate only to the extent that the "ruler" being used to make them can be calibrated to a perfect grid. On the scale of 0.01 μm no such grid exists.

Fortunately, absolute position accuracy of the order of 0.01 μm is not required for IC manufacture. The primary demand placed on lithographic systems is the ability to position one pattern level relative to another with a minimum registration error over the area of the circuit. Electrical techniques are most appropriate for the determination of such level-to-level errors on wafers. On the other hand, mask or reticle fabrication, which requires accurate pattern placement on a single pattern level, is best examined using interferometric measurement.

The measured pattern placement error at each site on the wafer can be composed of a variety of contributions, which depend on the type and status of the lithographic system. Scanners and steppers using global alignment are characterized by the set of grid errors illustrated in Fig. 5, although the physical mechanisms underlying the errors differ. In the case of the scanner, grid errors stem from global misalignment (translation and rotation), scan setting (scan and cross-scan distortion), and optical system performance (magnification and optical distortion). In the case of the stepper, the grid error represents the position of each field of the stepped pattern. In addition to global misalignment, stepping errors of the stage (nonorthogonality, scaling, and stage precision) are contributors. For the case of field-by-field alignment on the stepper, alignment accuracy and stage precision are the sole contributors to the grid error at each site.

Mathematical models exist for both grid and intrafield errors. The models must be tailored to the degrees of freedom of the particular class of scanner and/or stepper being considered. Previous authors [14–17], with

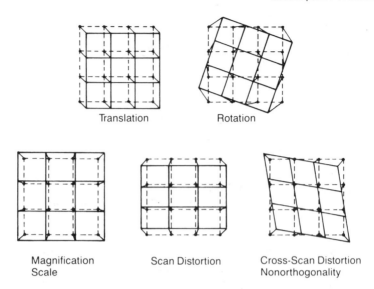

Translation   Rotation

Magnification   Scan Distortion   Cross-Scan Distortion
Scale   Nonorthogonality

Fig. 5.   Grid errors observed on both scanners and steppers.

the exception of Holbrook [18], have adopted decoupled scalar representations of the components of the pattern placement error. Here we adopt a more rigorous vector representation that ensures a direct correlation to physically meaningful lithographic parameters [19].

Under the assumption that the origin for grid errors (Δ_X, Δ_Y) is at the center of the wafer, a model for the grid error common to most scanners and steppers at each location (X, Y) can be written as:

$$\Delta = T_X \hat{\imath} + T_Y \hat{\jmath} + \Theta(-Y\hat{\imath} + X\hat{\jmath})$$
$$- \Phi(Y\hat{\imath} + X\hat{\jmath}) + S_X X\hat{\imath} + S_Y Y\hat{\jmath} \tag{4}$$

where $(\hat{\imath}, \hat{\jmath})$ are the unit vectors in the (X, Y) directions, (T_X, T_Y) are the translation components, Θ is the rotation coefficient, Φ is the orthogonality coefficient and (S_X, S_Y) are the scale coefficients. For a given system, these coefficients are determined by a least-squares fit of the Eq. (4) model to measured data.

The major difference between grid errors on the scanner and stepper is that the grid error of the scanner includes the distortion of the optical system, whereas that of the stepper includes the precision of the stage. Systematic errors observed in residuals, errors remaining after subtraction of those modeled by Eq. (4), will differ. A characteristic of the distortion of the ring-field optical system [3] used in the scanner, referred to as "Y bow," can be incorporated into the model by the term $B_X Y^2 \hat{\imath}$ added to the expression for Δ. On the other hand, stage precision errors tend to be random.

In the stepper case, grid errors tell only part of the story. Errors occurring within each field of the stepper image must be separately considered. The distinction between the intrafield and grid locations on the wafer is illustrated in Fig. 6. The intrafield errors can be broken down into components due to reticle misalignment, reticle position relative to the ideal object plane, and lens distortion (Fig. 7). Note that, while wafer misalignment is reflected in both the grid and intrafield errors, it is best analyzed as a grid error. For intrafield measurements, reticle misalignment can be distinguished from wafer misalignment by those components of translation and rotation that repeat from field to field.

Under the assumption that the origin for intrafield errors (δ_x, δ_y) is at the center of the field, the model for the intrafield errors can be written:

$$\boldsymbol{\delta} = t_x\hat{\mathbf{i}} + t_y\hat{\mathbf{j}} + \theta(-y\hat{\mathbf{i}} + x\hat{\mathbf{j}})$$
$$+ m(x\hat{\mathbf{i}} + y\hat{\mathbf{j}}) + (k_1 x + k_2 y)(x\hat{\mathbf{i}} + y\hat{\mathbf{j}})$$
$$+ d_3(x^2 + y^2)(x\hat{\mathbf{i}} + y\hat{\mathbf{j}}) + d_5(x^2 + y^2)^2(x\hat{\mathbf{i}} + y\hat{\mathbf{j}}) \qquad (5)$$

where $(\hat{\mathbf{i}}, \hat{\mathbf{j}})$ are now the unit vectors in the (x, y) directions, (t_x, t_y) are the translation components, θ is the rotation coefficient, m is the intrafield magnification or reduction coefficient, k_1 and k_2 are the trapezoid coefficients, and d_3 and d_5 are the coefficients of third- and fifth-order distortion of the lens.

The rotation, magnification (reduction), and trapezoid components are mechanically correctible by reticle repositioning. Distortion is an intrinsic

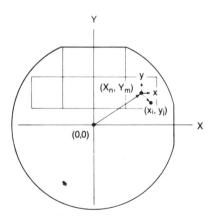

Fig. 6. On a stepper, each grid location (X_n, Y_m) on the wafer is surrounded by intrafield locations (x_i, y_j) over the area of a single field.

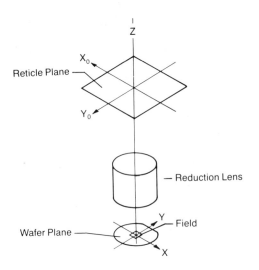

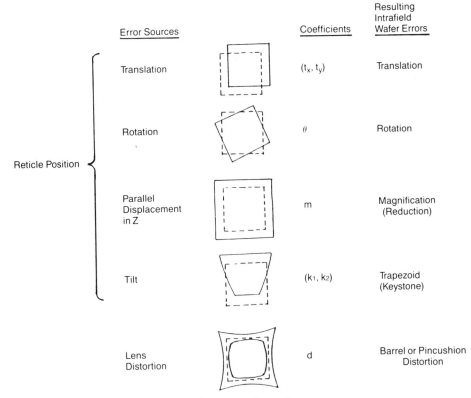

Fig. 7. Intrafield errors observed on steppers.

property of the stepper lens. In a good lens, the third- and fifth-order distortion components compensate for one another as nearly as possible across the field.

The distribution of the measured errors over the wafer is the key to understanding the overlay status of the lithographic system. For scanners, grid distributions alone need be considered. Stepper errors, however, are composed of both grid and intrafield errors. Depending on the system under test, the errors encountered may or may not be adjustable. The sheer number and complexity of the possible distributions underscore the need for the precise measurement and detailed analysis made possible by the electrical probe technique.

III. ELECTRICAL MEASUREMENT

A. Basics

As applied to lithographic characterization, electrical measurements consist of dc current and voltage determinations on patterns defined in conductive films on the wafer surface. In general, the resistance, the ratio of voltage to current, of an electrically conductive pathway, is inversely proportional to its cross-sectional area averaged along its length. When the pathway is defined in a film of uniform thickness, the inverse proportionally holds between the resistance R and the average width W,

$$R = \rho L / W, \tag{6}$$

where the proportionality factor is the product of the length L of the line and the sheet resistance ρ of the conductive film, in ohms per square ($\Omega/\square$). A simple interpretation of Eq. (6) is that the resistance R is the sheet resistance ρ times the number of squares L/W. For a line of length $L = 100 \ \mu m$ defined in a film of sheet resistance 50 $\Omega/\square$, a 2-μm linewidth would correspond to a 2.5-kΩ resistance, 50 squares of 50 Ω each connected in series.

We can rewrite Eq. (6) in the form

$$W = \rho L / R = \rho L G, \tag{7}$$

such that the linewidth can be deduced from the measured conductance ($G = I/R$), the "known" length of the line, and the sheet resistance of the film. Thus, we can determine the width of a microscopic line by an electrical measurement for which accurate and automated instrumentation exists [20].

B. Process

A simple process sequence that can be used to define an electrically probeable pattern is shown schematically in Fig. 8. A conductive film is deposited on an oxidized silicon wafer and coated with photoresist. The wafer is then exposed on the lithographic system under test, and the pattern is developed in the resist. The resist pattern is transferred to the conductive film by an etch process. The isotropy of the etch determines the fidelity with which the width at the base of the resist is replicated in the conductive film. Finally, the resist is stripped, leaving the conductive pattern on the insulating substrate. Previous authors have used variations on this process sequence that fit more naturally in their IC manufacturing facility [14,20,21]. Numerous approaches are possible, provided that the net result is an electrically isolated probeable pattern.

In order to understand the manner in which electrical measurements represent lithographic performance, it is informative to examine the profile of the etched conductive layer underlying the resist, as shown in the scanning electron microscope (SEM) photographs of Fig. 9. In both cases a nominal 1.1-μm line was exposed in ~ 1 μm of resist over a 1000-Å-thick chrome film. The difference in resist profiles is due to the fact that a postexposure bake (after exposure, before develop) was introduced to the process for Fig. 9b. A wet etch was used to pattern the chrome. The etch undercut is roughly equal to 0.15 μm on each edge, resulting in a linewidth narrowing of 0.3 μm relative to the base of the resist. Well-controlled film deposition and etch processes are required to ensure a constant undercut. Under these circumstances the change in the width of the chrome line accurately reflects the increase in resist linewidth due to the introduction of postexposure bake. However, the etched conductive pattern tells us noth-

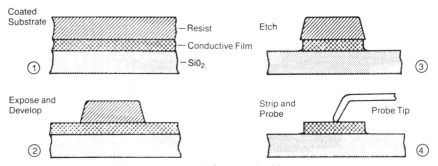

Fig. 8. Process to define a probeable pattern.

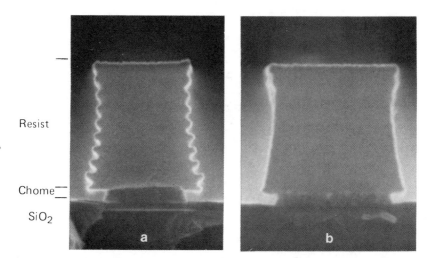

Resist

Chome

SiO$_2$

Fig. 9. Cross-sectional views of nominal 1.1-μm lines in 1.0 μm of AZ1470 resist on 1000 Å of wet-etched chrome. The process in (b) differed from that of (a) by the addition of a postexposure bake.

ing about the change in standing wave modulation evident in the SEM photographs. The observation of such profile effects continues to require the use of the SEM.

C. Probe

As illustrated in Fig. 8 and in the photograph of Fig. 10, one can perform the electrical measurement of the conductive pattern by bringing a set of probes in contact with the surface of the conductor. The configuration of the probes is fixed on a card mounted in the prober such that only the underlying patterned wafer is moved from site to site between measurements. While the contact area between each probe and the surface may be only of the order of a few square micrometers, relatively large pads ($\sim 100 \, \mu$m square) are required on the wafer pattern to accommodate probe stage inaccuracies.

Typical measurements involve the application of a constant drive current and the sensing of voltage on patterns such as those in Fig. 11. The simpler pattern of Fig. 11a allows for only two probes. The major problem with combining current drive and voltage sensing on the same probe is that

Fig. 10.   Probes in contact with a test wafer.

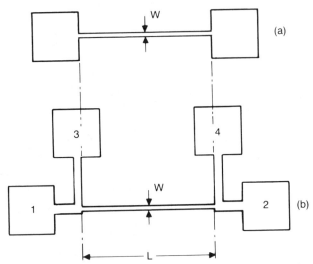

Fig. 11.   Linewidth test patterns. Pattern (a) requires common current drive and voltage measurement. Pattern (b) allows four-point, or "Kelvin," probing.

the probe-to-pad contact resistance is added to the resistance of the line we wish to measure. If voltage is measured at points separated from the current drive, as allowed by the pattern of Fig. 11b, the resistance of the line can be independently determined. To be specific, we drive current between pads 1 and 2 and measure voltage between pads 3 and 4. In this "full Kelvin" measurement no current flows through the voltage-sensing pads (for an ideal voltmeter). Consequently, there is no voltage drop across the contact resistance, and the voltage measurement is independent of the position of the probes relative to the line being measured.

Despite the insensitivity of Kelvin probing to contact resistance, good electrical contact is a must. For many conductive materials (e.g., chrome) an insulating oxide can form on the top of the conductive layer before probing. Such insulating barriers are typically thin (~ 100 Å) and can be broken down by a high-voltage pulse preceding each measurement. For example, 10 V across a barrier 100 Å thick corresponds to an electric field of 10^7 V/cm, greater than the electric field in a lightning bolt. In order to avoid possible damage to the narrow lines we wish to measure, the voltage prepulsing should be conducted through the wider interconnect lines, for example, pads 1–3 and 2–4 in Fig. 11b.

D. Sheet Resistance

In order to translate our conductance measurements into linewidth values, each measurement must be accompanied by a determination of the proportionality factor ρL. The length of line L is a constant determined by the mask- or reticle-making process. By keeping the ratio L/W large, typically ~ 100, we can ensure that possible variations in L will have a negligible impact on the electrical determination of W. The sheet resistance ρ, however, can vary significantly, both within a single wafer and from wafer to wafer. As shown in Fig. 12, local measurements of sheet resistance can be achieved in either of two ways: (1) measuring two lines of different width or (2) using a van der Pauw structure [14,22,23].

In the case where we measure two lines of equal length but differing width, Eq. (7) gives

$$\rho = \frac{1}{L} \frac{W_1 - W_2}{G_1 - G_2}. \tag{8}$$

The validity of Eq. (8) is based on the assumption that the difference between the two linewidths is a known constant independent of the absolute linewidths. The assumption holds true, provided that the linewidths chosen are imaged identically by the lithographic system; that is, they must

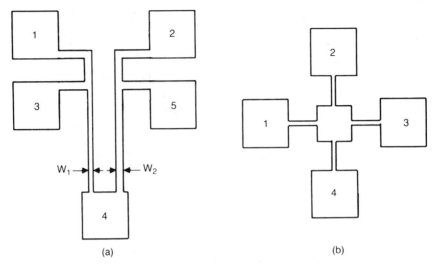

Fig. 12. Sheet resistance measurement patterns. (a) Two-linewidth pattern; (b) van der Pauw pattern.

be well within the system resolution. On the other hand, using the symmetric van der Pauw structure in Fig. 12, sheet resistance is given by

$$\rho = \frac{\pi}{\ln 2}\frac{V_{43}}{I_{12}}, \tag{9}$$

where V_{43} is the voltage sensed between pads 4 and 3 when current I_{12} is driving between pads 1 and 2. A more general expression, realized by averaging two measurements,

$$\rho = \frac{\pi}{2\ln 2}\left(\frac{V_{43}}{I_{12}} + \frac{V_{14}}{I_{23}}\right), \tag{10}$$

allows for possible distortions in the center square. While the van der Pauw approach is a widely accepted means of sheet resistance measurement, it should be noted that cases may exist where the van der Pauw voltages fall below the maximum precision range of the voltmeter being used. In such cases, the two-linewidth method of Eq. (8) becomes essential.

The electrical probe technique is very forgiving of the type of conductive films and the range of sheet resistances allowed. Almost any reasonably conductive layer is usable, including metals (NiCr, TiW, aluminum, chromium), doped layers (polysilicon, silicon), and silicides (Pt–Si, Ta–Si). Low-resistivity films can present a problem in that one must apply a higher drive current to obtain the same voltage drop as in the case of high-resistivity films. As noted above, the most difficult measurement is that of a van der Pauw pattern. For example, a 1-μm aluminum film might have $\rho =$ 0.03 $\Omega/\square$, in which case a drive current of 75 mA is required to produce a 500-μV drop. The risk one runs is that the I^2R heat can fry the connecting lines of the pattern if they are too narrow. Nonetheless, the proper choice of drive current and pattern design makes it possible to test films with values of sheet resistance ranging from 0.01 to 500 $\Omega/\square$ with no loss of precision.

Process capability is another factor that must be taken into account in the choice of material. Although measurement can compensate for gradual variations of sheet resistance over the wafer, local uniformity is required of the deposition and doping process. Any difference in sheet resistance at the position of the measured line relative to the position of the van der Pauw results in an error in the linewidth measurement. Furthermore, in order to avoid the need to distinguish resist image variations from etch variations, it is best to work with a well-controlled etch process. In the case of wet etching, best results are obtained using very thin (~ 300 Å) metal films (TiW, NiCr, Cr) in order to minimize the etch undercut relative to the measured linewidth. For dry processing, a high degree of anisotropy accompanied by high resist selectivity is desired.

E. Hardware

The essential hardware components of an electrical measurement system for characterizing lithography are stable current and voltage sources, accurate current and voltage meters, a prober, and automated means of connecting source or meter to the probe pins. While it is not difficult to home-brew a system, several commercial options exist. Until recently, the easiest approach was to use parametric test systems [24,25], which are designed primarily for the probe testing of the manufactured device and circuit characteristics. The versatility required of a parametric tester implies a switching matrix of relays in order to achieve any probe–source meter configuration. A system dedicated to lithographic characterization, such as the LithoMap system manufactured by Prometrix Corporation, allows the use of fixed channels, since the probe connections have been standardized. At the expense of flexibility, therefore, the LithoMap system realizes improvements in speed and reliability.

F. Precision

From the outset we have stated the need for routine measurement precision of 0.01 μm. At first blush this appears an impossible task. The secret to our success is that we measure a line that may be submicrometer in width but that is at least 100 μm in length. Sensitivity to overall width variations is leveraged by the length such that a small percentage of change in width results in a large change in conductivity. Similarly, susceptibility to local defects (nicks or bumps in the line) is suppressed by the ratio of their dimension to the length. Nonetheless, the achievement of the necessary precision requires drive current stability and voltage measurement accuracy of the order of 0.1%.

As we have already discussed, each linewidth determination requires two electrical measurements: conductance and sheet resistance. Therefore, we must consider the combined precision of the two in determining the ultimate linewidth precision. A histogram of more than 60 repeated linewidth determinations on a nominal 1.0-μm line in a 300-Å TiW film is shown in Fig. 13. The total range of measurements about the mean value of 0.82 μm is less than ±0.5%, or ±0.004 μm. (Note that this result is obtained despite the fact that the TiW film is damaged by the repeated probe impact.)

In most cases, the measurement precision is limited by the accuracy of the voltmeter. In order for a meter with 1-μV resolution to measure with 0.2% accuracy, the voltage must be 500 μV or greater. This limitation is most important for van der Pauw measurements where, for example, a

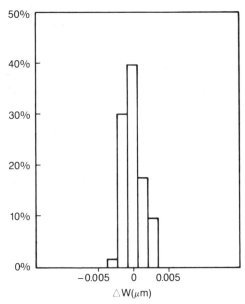

Fig. 13. Precision of electrical measurement as evidenced by the repeated probing of a nominal 1.0-μm line in a 300-Å TiW film.

current of 1 mA and a ρ of $1\mu/\square$ would give a measured voltage of only 221 μV. Thus, higher drive current is required on low-sheet-resistance films, with the caveat on current density noted in Section II.D.

On the other hand, narrower linewidths have higher resistance, which enables one to make more precise measurements. The same $\pm 0.5\%$ precision observed in Fig. 13, when applied to the measurement of a 0.1-μm line, would give a range of ± 5 Å. Are we really able to detect the presence of single atoms, as such sensitivity would suggest? Once again, we must recall that the measurement is averaged over a relatively long length of line. In the same sense that the measurement is not susceptible to nicks or bumps in the line, we do not detect single atoms. In fact, electrical measurement tells us nothing about the roughness of the line edges. At some limit, dependent on the conductive layer and the etch process, our implicit assumption of uniform linewidth breaks down. Beyond that limit the electrical linewidth measurement is more properly interpreted as an electrical continuity check.

G. Accuracy

Linewidth measurement precision does not lead directly to an accurate portrayal of the performance of a lithographic system. Our ability to determine accuracy in this sense is clouded by the interpolation of process steps — resist exposure, development, etch, etc. — between the aerial image of the lithographic system we wish to characterize and what we end up measuring on the wafer. Clearly, such accuracy is limited by the extent to which the processes are undertsood or controlled or both.

If we take a pragmatic view and accept the fact that we are characterizing the combination of the lithographic system with a given process, the accuracy issue assumes much more profound significance. A measurement is accurate to the extent that one is able to confirm it by independent and more accurate means. For example, we have compared electrical linewidth measurements with SEM measurements of the same chrome features (Fig. 14). In considering these results closely, however, we must keep in mind the fundamental differences between the two measurement techniques. The SEM measurements were taken from an image generated by secondary electrons over an approximately 10-μm length of line. The electrical results were a measure of the cross-sectional area of the chrome lines, as evidenced by their resistance to the flow of electrical current, averaged over a 30,000-μm meander length [21]. Which is the true linewidth? The most we can say is that, all things considered, the two techniques for measuring the so-called linewidth agree remarkably well.

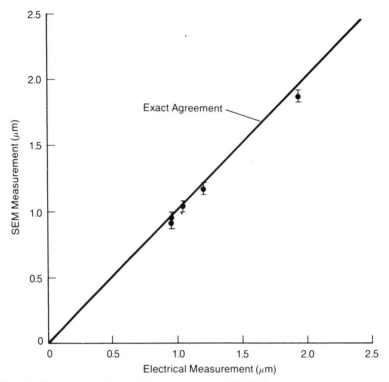

Fig. 14. Comparison of electrical measurements with SEM measurements over a range of linewidths.

The attempt to determine the truth of measurement brings to mind a vignette in Thoreau's *Walden* [25]:

> . . . the traveller asked the boy
> if the swamp before him had a hard bottom.
> The boy replied that it had. But presently
> the traveller's horse sank in up to the
> girths and he observed to the boy, I thought
> you said that this bog had a hard bottom.
> So it has, answered the latter, but you
> have not got half way to it yet.

We are still sinking. Yet we need not despair. As long as we adopt a standard technique for linewidth measurement, the problem of defining "linewidth" with absolute certainty pales in significance. As a measurement standard, electrical probing has much to recommend it: precision, speed, objectivity, and transportability (a volt is a volt the world over), to name a few of its fundamental attributes. The power and versatility for the imagery and overlay characterization of lithographic systems thus granted is demonstrated in the following sections.

IV. IMAGERY CHARACTERIZATION

A. Imagery Test Patterns

An example of an electrical test pattern appropriate to imagery characterization is shown in Fig. 15. The set of modules, each defined by a 2 × 8 layout of probe pads, contains nominal linewidths whose values are chosen on the basis of the approximate resolution limit for the lithographic system under test. Clearly, one wishes to have larger linewidths in the "safe" operating regime of the lithographic system, as well as smaller linewidths to test its resolving capability. Since the image quality can be a function not only of the advertised optical system parameters, but of the pattern location as well, the test patterns must be arranged in an array that samples the area of the printed field. The choice of an appropriate sample density is a trade-off among complete performance representation, pattern size, and test time.

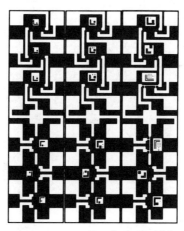

Fig. 15. Linewidth measurement test pattern including 0.6-, 0.7-, 0.8-, 0.9-, 1.0-, 1.1-, 1.2-, 1.5-, and 3.0-μm sizes in both the horizontal and vertical directions. The smaller features are not visible in the figure. Van der Pauw and visual test patterns are also present.

B. Uniformity

The evaluation of a scanner requires a distribution of test patterns over the full mask and wafer area. In order to isolate the imagery performance of a 1 : 1 system it is essential to map the linewidth variation over the test mask. This can be done by probing directly on the patterned chrome of the mask. Once the linewidth at each site on the mask is precisely known, the role of the 1 : 1 system in imaging each feature can be determined by taking the difference between the linewidth measured on the wafer and that of the corresponding site on the mask. A perfect exposure tool would produce an array of zeros.

Stepper characterization requires an array of test patterns over the reticle and printed field area. A reticle layout appropriate for testing imagery over

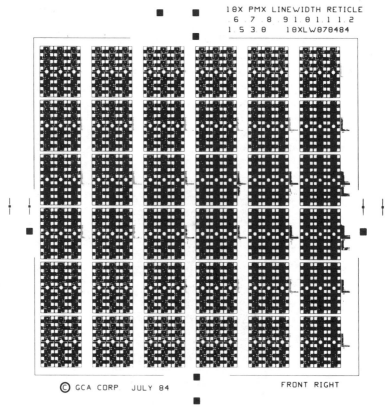

Fig. 16. Electrical test reticle for imagery evaluation of a 10 : 1 stepper consisting of a 6 × 6 array of linewidth measurement patterns.

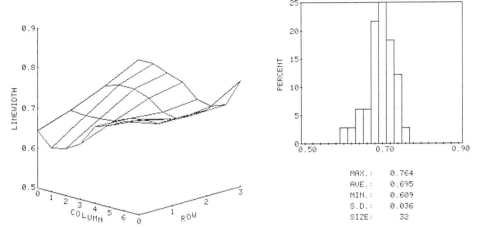

Fig. 17. Three-dimensional surface (a) and histogram (b) for the distribution of measurements of a nominal 1.0-μm line printed over the field of a 5:1 stepper (© Obelisk 1986).

the 10-mm-square field of a 10:1 stepper, which consists of a 6 × 6 array of the pattern of Fig. 15, is shown in Fig. 16. A reduction stepper does not require the same precision in characterizing the reticle as does the scanner in characterizing the mask, since the reticle features are 5 or 10 times larger than those printed on the wafer. A three-dimensional map and histogram of linewidth variation for nominal 1.0-μm lines over the 20-mm-diameter-field printed by a 5:1 stepper are shown in Fig. 17. It must be remembered that the wafer linewidth variation is the net result of illumination, reticle linewidth, lens performance, wafer flatness, and process variation over the field.

C. Exposure and Focus Variation

As we discussed in Section II, one can obtain complete characterization of the imaging quality of both scanner and stepper systems by printing series of fields or series of wafers (or both) over which system parameters, such as focus, exposure dose, and degree of partial coherence, are varied. The consequent variation of linewidth characterizes the system imagery performance.

Plots of the variation with exposure (at optimum focus) of nominal 1.5-μm and 1.0-μm features over the 20-mm-diameter field of a 5:1 stepper (NA = 0.30, λ = 436 nm) are shown in Fig. 18. The dashed curves show a theoretical match to the measured data averaged over the field, which is obtained using the modulation M as a free parameter in Eq. (3).

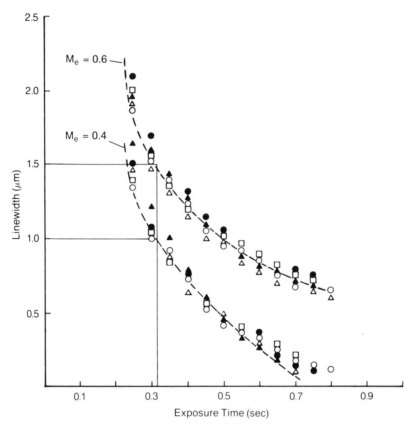

Fig. 18. Measured linewidth as a function of exposure for nominal 1.0- and 1.5-μm lines printed in 1.0 μm of AZ1470. The following legend refers to locations at the center and edge of the 20-mm-diameter field: O, Center; ▲, lower right; □, upper left; ●, upper right; △, lower left.

By separately fitting the data at each position in the field one could, in fact, determine the variation in modulation across the field. For the case shown, however, the variation does not appear to be significant.

By expanding our parameter space to encompass focus as well as exposure variation, we generate a family of curves, such as that shown in Fig. 19, analogous to the theoretical curves of Fig. 4. Here, the match of the theory of Eqs (2) and (3) to the experimental data, as shown in the dashed lines of Fig. 19, is obtained by a five-parameter fit (E_0, W_0, z_0, M_1, M_2), where M_1 is the focus-independent term in the modulation expression and M_2 is the coefficient of focus dependence. The agreement of the experimental results with the simple theory, illustrated in Figs. 18 and 19, substantiates our ability to characterize imagery performance.

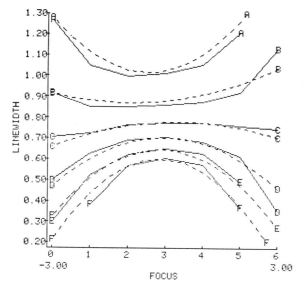

Fig. 19. Measured linewidth as a function of focus setting at different normalized exposure settings for a nominal 1.0-μm line. ———, Experiment; — — — —, theory.

V. OVERLAY CHARACTERIZATION

A. Overlay Test Patterns

The application of the electrical probe technique to the determination of overlay errors between two levels also makes use of its linewidth-measuring capability. All of the overlay test devices in widespread use today are modifications of the "stickman" pattern [26], which consists of a pair of lines whose widths are dependent on the misregistration at each site. As shown schematically in Fig. 20, by the performance of an alignment with the machine under test, the second-level "slot" pattern separates the wider first-level line into two parallel lines. Making use of Eq. (7), the misregis-

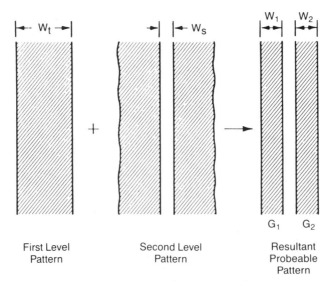

Fig. 20. Stickman pattern schematic.

tration Δx between levels is given by the relation

$$\Delta x = \rho \frac{L}{2}(G_1 - G_2) = \frac{W_1 + W_2}{2} \frac{G_1 - G_2}{G_1 + G_2}. \tag{11}$$

In the case of perfect overlay, the slot would be positioned precisely in the center of the first-level pattern such that $W_1 = W_2$ and $G_1 = G_2$. The same pattern rotated 90° is sensitive to the Y component of the overlay error. Using nominal linewidths of 5 μm on the probeable pattern, an error vector can be measured with a precision of 0.01 μm in both the X and Y directions.

An example of a pattern layout that includes a stickman structure among other vernier and optical test patterns is shown in Fig. 21. The layout maintains the 2 × 8 probe pad structure of the linewidth pattern in Fig. 15. A van der Pauw structure appears at the top of the pattern. For the stickman pattern appearing below the van der Pauw, the A level is in the form of a ring, which, when overlayed on the B level, overlaps the inner edges of the x- and y-oriented probeable lines to provide the equivalent of the stickman effect illustrated in Fig. 20. The design is such that the overlayed pattern reads "PROBE ME." Probe station operators will appreciate the value of this instruction for the proper identification of testable patterns on the wafer surface.

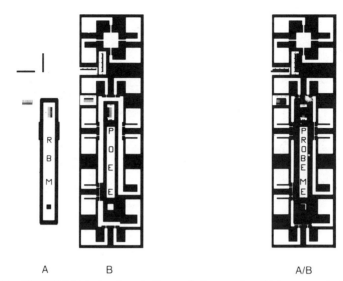

A   B   A/B

Fig. 21. PROBE ME test pattern before and after overlay. Stickman, optical verniers, box-in-box, and van der Pauw structures are present.

The simple stickman design has two disadvantages. The sheet resistance must be determined independently, for example, with the van der Pauw pattern in Fig. 21. More importantly, the simple pattern gives results that are sensitive to defects. For example, adhesion or etching problems that affect only one line can lead to false overlay measurements. Confidence in overlay specifications, which increasingly quote 2σ, 3σ, or total runout values, is undermined if even a small fraction of bad measurements pollutes the data set.

The redundant stickman pattern shown schematically in Fig. 22 eliminates these problems [15,28]. The design consists of two simple stickman patterns in which the second-level pair of slots are offset from one another by a known distance Δ_0. Linewidths are chosen that are well within the resolution of the lithographic system under test, such that the relation $W_1 + W_2 = W_3 + W_4$ holds true independent of focus and exposure variations. By applying Eq. (7), we obtain

$$\rho L = \frac{2\Delta_0}{(G_1 + G_3) - (G_2 + G_4)}, \tag{12a}$$

$$\Delta x = \frac{\Delta_0}{2} \frac{G_1 G_3 - G_2 G_4}{G_1 G_4 - G_2 G_3}. \tag{12b}$$

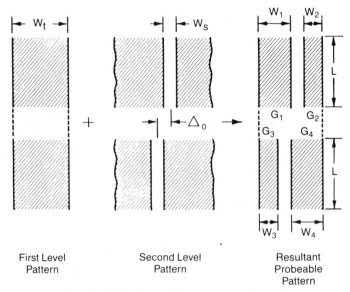

Fig. 22. Redundant stickman pattern schematic.

Only the known offset Δ_0 and the four measured conductances appear on the right-hand side of Eq. (12). Thus, the misregistration Δx is determined independently of the site-to-site sheet resistance and lithography variations. In effect, we are making use of the two-linewidth technique of sheet resistance determination discussed in Section II.D.

A layout of the redundant stickman test structure is shown in Fig. 23. Overlay of the A-on-B-level pattern, which reads "PROBE ME II," results in eight probeable lines (four in x and four in y, corresponding to the lines shown in Fig. 22). Current is driven from one end of the meandering structure to the other, and voltages are picked off along the way. One then determines the registration error by substituting the measured conductance values in Eq. (12). The integration of the sheet resistance measurement in the test structure reduces both the required pattern size and the probe time.

The redundancy of the pattern in Fig. 23 also allows a simple test for defective patterns. From the fact that the relation $G_1 + G_2 = G_3 + G_4$ should hold independent of overlay error, we can apply a pass–fail criterion,

$$\left| \frac{G_1 + G_2 - (G_3 + G_4)}{G_1 + G_2 + G_3 + G_4} \right| \le \epsilon, \tag{13}$$

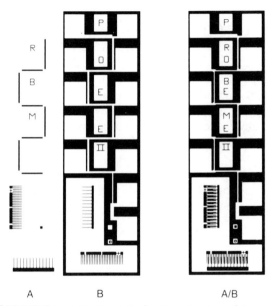

A   B   A/B

Fig. 23. PROBE ME II test pattern, employing the redundant stickman structure, before and after overlay.

above which the measurement is deemed erroneous. The value of ϵ, typically ~0.01, can be set during data acquisition. Thus, it provides an objective means of accepting, repeating, or rejecting the measurement.

The use of the data-weeding technique makes the overlay measurements impervious to processing variations. On the other hand, the yield of good

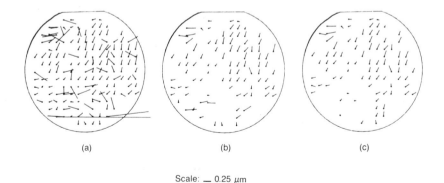

(a)   (b)   (c)

Scale: — 0.25 μm

Fig. 24. Weeding example on a poorly processed wafer using the redundant stickman pattern. As the erroneous data points are removed, the fundamental characteristics of the scanner emerge. (a) $\epsilon = 0.5$; 150 points. (b) $\epsilon = 0.05$; 91 points. (c) $\epsilon = 0.01$; 84 points.

data points as a function of ϵ is indicative of process quality or measurement precision or both. The wafer vector maps in Fig. 24, obtained from a very poorly processed wafer, provide graphic proof of the utility of the weeding technique.

The use of redundancy points out a distinction that must be drawn between the ability to make accurate linewidth measurements, as discussed in Section II.G, and the accuracy of overlay measurements. In the case of overlay we are always determining the difference between two linewidths. To the extent that the two are imaged and processed identically, therefore, the accuracy of the overlay determination is insensitive to imagery and process effects. In other words, the dependence of overlay accuracy on image and process control is a second-order effect that can be suppressed by the use of the weeding factor.

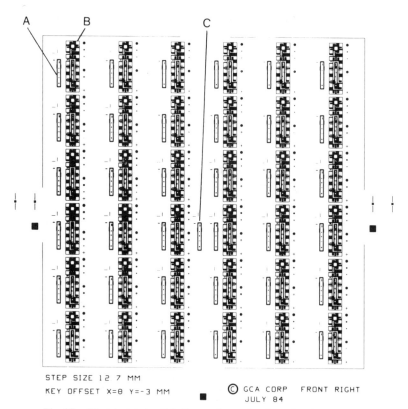

STEP SIZE 12 7 MM
KEY OFFSET X=8 Y=-3 MM   © GCA CORP FRONT RIGHT
JULY 84

Fig. 25. Electrical test reticle for overlay evaluation of a 10:1 stepper.

B. Overlay Test Reticle

A wide variety of ingenious test pattern layouts for the characterization of the overlay performance of lithographic systems have been developed. As an example, a new reticle appropriate for electrical overlay testing of the GCA stepper is shown in Fig. 25. The design consists of an array of PROBE ME patterns (see Fig. 21). The layout shown is that of a 10× reticle containing a 6 × 6 array. The corresponding 5× version of the PROBE ME reticle contains a 12 × 12 array. Near the center of the field appears a single second-level half of the PROBE ME pattern, denoted C in Fig. 25.

The layout of the PROBE ME reticle is unique in that it makes possible numerous different modes of use. For intrafield error measurement, two distinct applications exist. On one hand, the second-level C pattern can be positioned and printed over each first-level B pattern in the array in a series of blind steps in order to determine the intrafield errors present on a single machine relative to the grid of the interferometric stage. On the other hand, to measure the intrafield error of one machine relative to another, the full field of the reticle can be printed on one machine and then printed again on a second machine shifted by 420 μm to allow the matching of the B and A patterns at each position in the field. A similar procedure carried out on a single machine enables one to study overlay stability over the time elapsed between exposures.

For grid error measurement, blind stepping of the B-on-A array on a single machine gives a measure of stage precision, while the interlocking of adjacent fields can determine orthogonality. Finally, the presence of local and global alignment targets at each point in the reticle array allows any number of intrafield, interfield, or full wafer alignment test scenarios.

C. Overlay Measurement Applications

Exhaustive studies of overlay performance are made possible by the electrical probe technique. Two rudimentary examples are considered here.

In the case of grid errors, we record the position on the wafer (X, Y) and the misregistration components $(\Delta X, \Delta Y)$. The distribution of errors over a simple wafer can be displayed graphically in the form of a full wafer vector map, such as that shown in Fig. 26a. For the case illustrated, the first-level pattern was contact-printed and the second level was printed by a scanner. We make the assumption that the measured grid error at each point is composed of the misalignment, magnification, and distortion terms represented by Eq. (4). The measured data are least-squares-fitted to

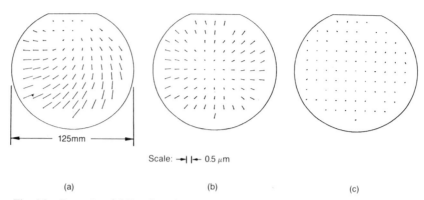

Scale: →| |← 0.5 μm

(a)   (b)   (c)

Fig. 26. Example of full-wafer grid error measurement and analysis of a 1:1 scanner relative to a contact print over a 125-mm-diameter wafer. (a) Measured; (b) alignment removed; (c) magnification removed.

the six-parameter model. Figure 26b shows the effect of subtracting the alignment errors (translation and rotation) from the measured data of 26a. Subsequent removal of the magnification error results in the map of Fig. 26c. The vectors in Fig. 26c are all below 0.15 μm in magnitude. Our instant picture from the particular wafer tested is of a system with good distortion characteristics whose overlay performance is dominated by alignment and magnification errors.

In the case of intrafield errors, we record the position within the field (x, y) and the misregistraton components $(\delta x, \delta y)$. The distribution of errors over a single field can be displayed graphically in the form of vector maps, such as those shown in Fig. 27. For the case illustrated, the first and second levels were printed by two different 5:1 steppers. We make the assumption that the measured intrafield error at each point is composed of the misalignment, platen positioning, and lens distortion terms illustrated in Fig. 7 and represented mathematically by Eq. (5). The measured data are least-squares-fitted to the model.

The residuals are the errors left over after the modeled components have been removed. The fact that the residuals are more or less random is testament to the capacity of the model to identify the systematic components in the measured data.

As in the case of the wafer map of the scanner, the intrafield map of the stepper gives an immediate view of the system status. In neither case are the data meant to be representative. Rather, they illustrate the fundamentals of the measurement and analysis approach. Complete characterization of any system requires many such maps, averages, and histograms to establish the form and extent of the error distributions.

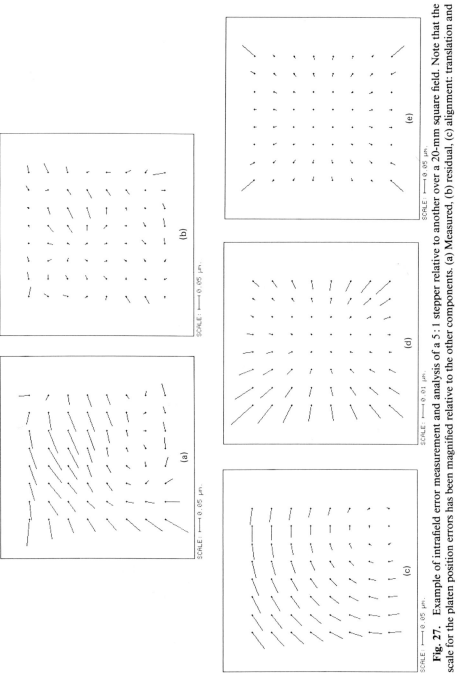

SCALE: ⊢——⊣ 0.05 μm.

(a)

SCALE: ⊢——⊣ 0.05 μm.

(b)

SCALE: ⊢——⊣ 0.05 μm.

(c)

SCALE: ⊢——⊣ 0.01 μm.

(d)

SCALE: ⊢——⊣ 0.05 μm.

(e)

Fig. 27. Example of intrafield error measurement and analysis of a 5 : 1 stepper relative to another over a 20-mm square field. Note that the scale for the platen position errors has been magnified relative to the other components. (a) Measured, (b) residual, (c) alignment: translation and rotation, (d) platen position: reduction and trapezoid, (e) lens distortion.

VII. CONCLUSION

There is little doubt that electrical measurement — by virtue of its precision, speed, objectivity, and transportability — will assume a prominent position, alongside optical and electron microscopy and laser interferometry, in the pantheon of lithographic metrology tools. This is not to say that electrical measurement comes without cost. As we have noted throughout, the superior precision and data-handling capacity of the electrical probe technique is achieved by an investment of effort in the design of appropriate test patterns and processes coupled to the automation of data gathering and analysis.

Over the next few years we can expect to see an increasing reliance on electrical measurement for characterizing lithography. In addition to its application to imagery and overlay evaluation, which has been discussed here, electrical probing will play a role in defect detection and yield analysis through the use of interdigitated test structures [29–31].

While the variety of metrology tools mentioned here are available for off-line lithographic characterization, the evolution toward automated IC manufacture indicates the ever increasing need for on-line metrology and control. The process limitations of electrical measurement do not recommend it as an on-line technique. Likewise, the sluggishness of existing microscopy and interferometric tools argues against their use.

Advancements in stepper lithography have initiated an interesting trend. The very properties that define a state of the art stepper — high-resolution imagery (whether it be optical, electron-beam, or x-ray), precise local focus and alignment capability, and high-speed interferometric stages — are the makings of an on-line metrology tool. We can envision a future in which the stepper can monitor and control its own lithographic performance.

Much of what we have learned is universally applicable to lithographic characterization, regardless of the measurement technique. We began our discussion with Faust's lament on the futility of knowledge. Yet in the end, Faust is driven to know

> That I may detect the inmost force
> Which binds the world, and guides its course;
> Its germs, productive powers explore
> And rummage in empty words no more.

It is the germs and productive powers of today's lithographic systems that we seek in order to replace the empty words of "specsmanship" with a true understanding of performance. We anticipate that in the long run our understanding of lithography, faciliated by the application of electrical measurement, will pave the way for the development of still more productive lithographic systems.

APPENDIX A

The light energy variation at the mask plane (see Fig. 1) of an incoherently illuminated, periodic mask pattern can be expressed as a Fourier series of spatial frequencies nv_0, where v_0 is the fundamental frequency of the pattern:

$$E_M(x) = A_0 - \sum_{n=1}^{\infty} A_n \cos(2\pi n v_0 x). \qquad (A.1)$$

The coefficients $A_0, A_1, \ldots, A_n$ are the Fourier coefficients of this pattern.

The imaging system acts as a low-pass spatial filter whose characteristic is given by the modulation transfer function curve in Fig. 3. Thus, the spatial variation of the exposure energy in the aerial image $E_A(x)$ is

$$E_A(x) = A_0 - \sum_{n=1}^{\infty} M(nv_0, z)A_n \cos(2\pi n v_0 x), \qquad (A.2)$$

where $M(nv_0, z)$ is the modulation of the optical system at frequency nv_0 and focal position z.

If we assume that the optical system modulation coupled to the photoresist sensitivity is sufficient only to image the fundamental frequency v_0, Eq. (A.2) simplifies to

$$E_A(x) = E\,[1 - (4/\pi)M(v_0, z) \cos(2\pi v_0 x)] \qquad (A.3)$$

where E is the average energy in the sinusoidal image.

For the case of an equal line and space pattern ($v_0 = 1/2W_0$) imaged in photoresist characterized by an energy threshold E_0, we can determine the printed linewidth from the position at which the aerial image energy equals the resist threshold energy:

$$E_A\left(\frac{W}{2}\right) = E\left[1 - (4/\pi)M(W_0, z) \cos\left(\frac{\pi}{2}\frac{W}{W_0}\right)\right]$$
$$= E_0. \qquad (A.4)$$

Consequently, we obtain Eq. (2).

ACKNOWLEDGMENT

This work could not have been completed without the stimulus of many discussions with my colleagues S. Cheng, T. A. Brunner, and R. V. Tan. The SEM photographs of Fig. 9 are the work of G. Ketley. The patterns of Figs. 15, 16, 21, 23, and 25 were developed in collaboration with the maestros of GCA reticle design, P. Reynolds and C. Sager. The computer graphics representations of Figs. 17, 19, and 27 were generated on the MONO-LITH™ lithography workstation produced by Obelisk, Inc.

REFERENCES

1. J. W. Von Goethe, "Faust: A Tragedy" (B. Taylor, transl.). Warne, New York, 1889.
2. D. A. Markle, *Solid State Technol.* **17**, 50 (1974).
3. J. D. Cuthbert, *Solid State Technol.* **20**, 59 (1977).
4. M. C. King, *IEEE Trans. Electron Devices* **ED-26**, 711 (1979).
5. J. Roussel, *SPIE Semin. Proc.* **135**, 30 (1978).
6. W. C. Schneider, *SPIE Semin. Proc.* **174**, 6 (1979).
7. G. L. Resor and A. C. Tobey, *Solid State Technol.* **22**, 101 (1979).
8. M. Born and E. Wolf, "Principles of Optics," 4th Ed., pp. 419, 441. Pergamon, Oxford, 1970.
9. H. Moritz, *IEEE Trans. Electron Devices* **ED-26**, 705 (1979).
10. M. C. King, *in* "VLSI Electronics: Microstructure Science" (N. G. Einspruch, ed.), Vol. 1, p. 41. Academic Press, New York, 1981.
11. A. Offner, *Photogr. Sci. Eng.* **23**, 374 (1979).
12. B. J. Lin, *IEEE Trans. Electron Devices* **ED-27**, 931 (1980).
13. A. R. Neureuther, P. K. Jain, and W. G. Oldham, *SPIE Semin. Proc.* **275**, 110 (1981).
14. D. S. Perloff, *IEEE J. Solid-State Circuits* **SC-13**, 436 (1978).
15. C. P. Ausschnitt, T. A. Brunner, and S. C. Yang, *SPIE Semin. Proc.* **334**, 17 (1982).
16. W. H. Arnold, *SPIE Semin. Proc.* **394**, 87 (1983).
17. D. MacMillen and W. D. Ryden, *SPIE Semin. Proc.* **334**, 81 (1982).
18. D. Holbrook, *Proc. Kodak Microlithogr. Conf., San Diego, Calif.* (1983).
19. R. V. Tan and C. P. Ausschnitt, *SPIE Micron and Submicron Metrology Conf. 565, San Diego, California, 1985.*
20. M. G. Beuhler, M. G. Grant, and W. R. Thurber, *J. Electrochem. Soc.* **125**, 650 (1958).
21. R. Allen *et al., SPIE Semin Proc.* **470**, 111 (1984).
22. D. S. Perloff, F. E. Wahl, and J. Conragen, *J. Electrochem. Soc.* **124**, 582 (1977).
23. L. J. Van der Pauw, *Philips Res. Rep.* **13**, 1 (1958).
24. C. Chrones, *Semicond. Int.* **3**, 113 (1980).
25. J. S. Howard and J. Nahourai, *Solid State Technol.* **21**, 48 (1978).
26. H. D. Thoreau, "Walden," p. 330. Princeton Univ. Press, Princeton, New Jersey, 1973.
27. I. J. Stemp, K. H. Nicholas, and H. E. Brockman, *IEEE Trans. Electron Devices* **ED-26**, 729 (1979).
28. C. P. Ausschnitt, T. A. Brunner, and S. C. Yang, *SPIE Semin. Proc.* **342**, 65 (1982).
29. J. H. Bruning, *Symp. Electron, Ion Photon Beam Technol., 15th 1979.*
30. T. A. Brunner, C. P. Ausschnitt and D. Duly, *Solid State Technol.* **26**, 135 (1983).
31. C. Mallory *et al., Solid State Technol.* Nov. (1983).

Index